EVOLUTION: THE HIDDEN ASSUMPTIONS

HOW A SIMPLE IDEA OF SELECTIVE BREEDING
BECAME THE THEORY OF EVOLUTION

All rights reserved © Amitabha Lahiri 2018

Disclaimer
No part of this book may be reproduced or transmitted in any form or by any means, mechanical or electronic including photocopying or recording or by an information storage and retrieval system or transmitted by e-mail without permission from the author.
This book is not to disregard any alternative views on the subject,
but a personal analysis and interpretation of available literature.
The views are those of the author alone
Published by Amitabha Lahiri. Singapore

Email : novumlibrum@gmail.com

Cover art and design by Sanjana Lahiri

Contents

- the Theory Of Evolution: Should There Even Be A Discussion Today? 5
- The Birth Of A Theory 9
- Darwin's Big Idea 12
- Understanding Evolution By Means Of Natural Selection 16
- The Modern Theory Of Evolution 18
- Smooth Transitions Or Sudden Jumps 21
- A Closer Look At The Word "Evolution" 29
  - Ghost Cars 33
- Levels Of Biological Transformation: A New Model For Looking At Evolution 37
- The Conditions For Evolution By Natural Selection 47
  - What Does It Take For An Animal To Be Viable? Just Perfect Form, And Function 49
  - The Making Of A Living Animal 55
  - Making Of The Reproductive System 59
- The Absolute Conditions For Evolution: The Bio-Feasibility Model 61
- Testing Biological Feasibility Of The Three Levels Of Transformations 64
  - The Biological Feasibility Of Alpha Transformation (Emergence Of Multicellular Animals From One Celled Animals) 65
  - Beta Transformation: Evolution Of Classes (Mammals From Reptiles) 71
  - The 250 Million Year Periscope 75
  - Gamma Transformation: Emergence Of New Forms Within Mammals (Origin Of Orders) 79

What Of The Scientific Evidence: Where Does It Fit? 82
    Computer Modelling: The Blind Watchmaker: An Erroneous Paradigm? 86

How Does Biological Feasibility Model Compare With Scientific Observations.. 88

Darwin's Assumptions 94

Laying Down Rules Around Darwinism 101

Summary.. 106

Points To Ponder 110
    Each Era Had An Entire Ecosystem Of Animals 110
    Absence Of Imperfect Forms 111
    A Series Of Fortunate Events, Or Planned 'Terraforming' Of Earth?... 113
        About The Author 116
        Acknowledgements 117

THE THEORY OF EVOLUTION: SHOULD THERE EVEN BE A DISCUSSION TODAY?

Today we know the theory evolution for fact. It is taught in every science curriculum, and is accepted as the sole scientific explanation for the existence of all life on earth.
 But is it the absolute scientific truth? Have we actually answered all the questions about life on this planet or we have conveniently accepted the theory for lack of a better one?
 Let us examine!
The story of evolution as narrated by science is as follows:

Life seems to have appeared through a lucky accident when Earth was a hot swirling pool of chemicals. Within this boiling pool, life appeared through a lucky accident (scientists have no clue how, but there are several theories). The very first life that appeared about four billion years ago, was a group of bacteria. They were very special, they knew exactly how to process these toxic chemicals and release oxygen and water. They created very first oceans on earth and oxygenated the atmosphere, the event is known as the 'great oxygenation event'.
 Water and oxygen prepared our planet to harbour life as we know it. Somewhere around two billion years ago one-celled animals and plants (the eukaryotes, the ones with a true nucleus) appeared in the water. They possibly evolved from the bacterial ancestors, but again, no one really knows about how this transformation actually occurred, but again there are several theories to fill the gap.
Roughly 5oo million years ago, another mysterious event took place , these one cell organisms made a huge leap, they transformed into multicellular animals, such as sponges, worms, arthropods (insects), molluscs (nautilus and squids), and early ancestors of vertebrates (animals with back bones) to name a few.
The small vertebrate ancestors evolved into fishes, which then developed limbs to exploit the swamps as the amphibians like salamanders. The amphibians couldn't leave the swamps. They were tied to water because their skin had to be wet all the time

and their jelly-like eggs could only survive in water. They 'evolved' and gave rise to reptiles that had scaly skin and hard-shelled eggs, and were capable of inhabiting dry land.
Some of these reptiles evolved to grow feathers and gave rise to birds.

Some other reptiles, through a yet unknown mechanism developed a uterus and began to give birth to live offspring and became the mammals.
Mammals proliferated to give rise to several orders including apes. Strangely again by some quirk of nature, the apes were the only group that evolved brains capable of abstract thought, leading on to early humans and then to human civilisation.

There are a few peculiarities in this story. One may notice, for a scientific narrative there are too many coincidences and happy accidents.

Every key step, from the origin of life, to the emergence of multicellular life forms, and eventually transformation vertebrates leading to appearance of mammals is wrapped in mystery and held up with theory rather than hard evidence

At the same time, if we take an unbiased look, the unfathomable complexity that living organisms have, makes one wonder whether the assumption that everything happened through a chain of random events is actually true.

Several intellectuals in the past have noticed these incongruities.

In 1802, **William Paley** published an interesting text called 'Natural Theology; or, Evidences of the Existence and Attributes of the Deity'. Interesting, because he was one of the first individuals who recognised the presence of sophisticated design in human anatomy.
He noticed the body had moving parts that that fit elegantly with one another and worked together, just like a complex machine! He wrote *"the necessity, in each particular case, of*

an intelligent designing mind for the contriving and determining of the forms which organized bodies bear".

His most famous analogy is the watchmaker analogy. He argued that when you see a complex mechanical structure such as a watch, it prompts you to think of a watchmaker. Similarly, complex anatomical structures should prompt the thought of a creator!
This was probably the earliest thought of perceiving living creatures as sophisticated machines.

That though has not gone extinct. Today, the concept of *Intelligent Design* upholds this idea that living organisms demonstrate such an extraordinary complexity in form and function, it is unlikely that they are a product of a random process. They are more likely to be products of a higher intelligence.

One of the strongest arguments in favour of *intelligent design* was proposed by **Michael Behe**. He coined the term "irreducible complexity" in his book "Darwin's Black Box", He defined irreducible complexity as *"a single system which is composed of several well-matched interacting parts that contribute to the basic function, wherein the removal of any one of the parts causes the system to effectively cease functioning"*
Dr Behe studied the tail like cilium if bacteria. He found that the structure at the base was a tiny motor constructed with molecules that rotated to help bacteria swim. It had parts that functioned together to perform a single function, like parts of a machine. This observation was a strong argument against the gradual incremental process of evolution because in absence of all these parts together at the same time the cilium would be non-functional.
The concept of *intelligent design* has been labelled 'pseudoscience' by the scientific community. Interestingly, there is an entire branch of science today that is dedicated to the study of 'design' in nature, and the incorporation of these **biomimetic designs** to enhance the efficiency of machines.

Evolutionary scientists use the term ***body- plan*** to describe the unique ***design*** of animals in individual phyla.
Evolutionists are well aware of the fact that there are several anomalies in the available evidence that contradict the idea of evolution, yet the discussion on these findings remains a side issue in scientific circles.

This persisting dichotomy on the matter   mandates a fresh debate.

THE BIRTH OF A THEORY

How did the theory of evolution come about?

Since the dawn of time, humans had always acknowledged the presence of a higher power. Rain, thunder, fire, famine everything that was beyond human comprehension was attributed to this power.

With the passage of time, humans began to think. Some humans had begun to rationalise the world that surrounded them and started looking for logical explanations for natural occurrences that surrounded them. They started looking for explanations that did not involve supernatural interventions. They had begun to detach the physical world from the celestial world.

One of the first works to appear was Lucretius's De *Rerum Natura* (On the Nature of Things) in the first century BCE. He was probably one of the earliest philosophers to propose that the universe operated according to physical principles, and not divine intervention. Lucretius described in poetry, the principles of atomism; the nature of the mind and soul, the development of the world and its phenomena.

Slowly but steadily, mankind was discovering physical principles behind the fire, the electricity and the rain, and began to construct devices that could harness these elements.

As the time went by, some intellectuals turned their attention towards the world of living creatures.

The fundamental question was, where did these living creatures come from? And the answer had to be an earthly and logical, devoid of celestial intervention.

Various works of scientific thought began to appear in the western world.

In 1735 Carl Linnaeus classified living organisms in a hierarchical nature, placing living organisms in the order of complexity. He gave birth to the field of *taxonomy* the science of classification and naming living creatures.

His idea that living organisms ranged from simple, such as sponges to complex such as apes, probably set the stage for evolutionary thinking that would happen almost a century later.

The concept that animals could *transform (and hence evolve)* came in 1809, from another visionary in science, Jean-Baptiste Lamarck. His "transmutation" theory, envisaged the process of 'spontaneous generation' (living creatures arising from inanimate matter) continually producing simple forms of life, that developed greater complexity in parallel lineages, and these lineages adapted to the environment by inheriting physical changes that resulted from use or disuse of body parts by the parents.

Fifty years later, in 1859 a remarkable manuscript appeared: *'On the origin of species by means of natural selection'* that marked a turning point on how we viewed the living world. It was authored by a naturalist, Charles Darwin.

He suggested, that new species originated through the process of *'natural selection'*, of the members of the population who were best suited to win the *struggle for existence*. Over millennia, these winning traits ('good' variations) accumulated such that the offspring no longer seemed to resemble their ancestors and came to be a new species

The theory was the answer the scientific community was waiting for! Science now had the perfect formula that could explain the emergence of every life form of Earth. It made sense, appealed to intellect and it could even be demonstrated in real life. The problem solved, the scientific world then proceeded to collect evidence and bolster the theory until it became a fact.

Arguments against evolution were given the colour of a 'religion versus science' debate. It became the debate between creationists, who believe in a mysterious entity, about whom references are found in archaic texts, and the scientific minded who live by the evidence demonstrated clearly by science.

Following its dramatic victory over creationism, the theory of evolution then became a staple in scientific curricula, from primary school to highest levels of academia until it came to be regarded one of the greatest insights in the history of science itself.

To understand the power of this idea one must understand the idea itself.

DARWIN'S BIG IDEA

The manuscript that laid the foundation for the theory of evolution as we know it was called *'On the Origin of Species by Means of Natural Selection or, the Preservation of Favoured Races in the Struggle for Life'* or simply the 'Origin of Species'. It was published on 24 November 1859.
It was authored by Charles Darwin.
Darwin was a born naturalist. He had a curious mind and was fascinated by the natural world. Pondering over the distribution of wildlife and fossils he had collected on his five-year voyage aboard the *Beagle*, he conceived his theory of natural selection.
He had noticed that animals were born with variations, i.e. they were never exact replicas of their parents. Animal breeders could choose these individuals with variations and selectively breed them to mould their physical features and produce new breeds.
These effects could be easily observed in animals, such as horses, sheep, birds, dogs. Sometimes the effects of selective breeding were so profound, that the offspring barely resembled its ancestors.
He himself was a pigeon fancier. He observed that he could select pairs of birds with certain features and produce offspring with different shapes of beaks, feathers, and colours. Fascinated by these observations he wrote, *"The diversity of the breeds is something astonishing, yet all were descended from one species of rock pigeon."*
He called this phenomenon *'variation under domestication'*
He believed that in nature, species may change in the same way as they did under domestication. The process of selective breeding occurred in nature as well, only that the selection was based on the ability of the members of the species to survive. In other words, the successful survivors of a species had the opportunity to breed and produce an even better offspring while the weaker members would perish.
He wrote *"Owing to this struggle for life, any variation, however slight and from whatever cause proceeding, if it be in any degree profitable to an individual of any species, in its infinitely*

*complex relations to other organic beings and to external nature, will tend to the preservation of that individual, and will generally be inherited by its offspring ... I have called this principle, by which each slight variation, if useful, is preserved, by the term of **Natural Selection,** in order to mark its relation to **man's power of selection".***

The idea had stemmed from the theory proposed by Malthus in 1798, almost fifty years earlier.

In his writing 'An Essay on the Principle of Population', Malthus observed that *"an increase in a nation's food production improved the well-being of the populace, but the improvement was temporary because it led to population growth, which in turn restored the original per capita production level".*

Darwin was convinced that animal populations behaved in that same way. They bred until populations exceeded the availability of resources.

As the resources dwindled the members of the species would compete such that the fittest would survive to produce offspring. The offspring would be endowed with variations advantageous to their survival, gradually leading to the accumulation of such variations over generations. The result may be such that over thousands of years, would lead to the emergence of new species.

However, there were some obvious problems with his idea.

First if one were to believe that to be true one would expect a seamless continuum of intermediate forms between different species, which we do not. It means that if we have an ancestral species and a new species there should literally be thousands of intermediate species between the two. For example, if we consider dogs to have descended from wolves, we should see thousands of species representing the spectrum intermediates between the two. But we don't.

Darwin had noticed this problem. He wrote a chapter addressing this problem.

The chapter was called 'the difficulties in the theory'

He pondered

"why, if species have descended from other species by insensibly fine gradations, do we not everywhere see innumerable transitional forms? Why is not all nature in confusion, instead of the species being, as we see them, well defined?"

Darwin explained this by hypothesising that *"the competition between different forms, combined with the small number of individuals of intermediate forms, often lead to the extinction of such forms"*

Another important problem that he needed to address was the evolution and development of internal organs. He had observed that internal organs were different in different groups of animals. The differences were not just in sizes of these organs but, the organs performing the same function were structurally distinct.
In order to suggest that all animals had common ancestry one had to explain the transformation of internal organs as well.

He was so convinced of the robustness of his theory and concluded that the development of internal organs also followed the same principles as did the external features of animals.
He wrote:

"If it could be demonstrated that any complex organ existed, which could not possibly have been formed by numerous, successive, slight modifications, my theory would absolutely break down. But I can find out no such case".

And with these arguments in place the theory was complete.

Darwin's theory could be summarised as:

.
Offspring are not identical to their parents but have heritable variations.
As populations grow, individuals compete for resources *(struggle for existence)*

Offspring with advantageous variations are more likely to survive the struggle and, breed successfully, spreading their variations in the population *(natural selection)*.

Over a long period of time, these variations accumulate such that the offspring no longer resemble their ancestors, thus becoming a new species.

This can be summarised as *origin of species by means of natural selection.*

UNDERSTANDING EVOLUTION BY MEANS OF NATURAL SELECTION

Essentially the concept of evolution can be illustrated as:

We have a population A, Population A has variants: called the B variants .B variants are hardier than the A's. Gradually the B variants survive and have greater opportunity to bear offspring, which possess their hardiness, gradually the entire population is replaced by the B variants.

To illustrate this phenomenon we can use one of the most iconic examples of evolution the Peppered Moth.

The phenomenon was noted in the United Kingdom during the Industrial Revolution. The common species of Peppered Moth was the white-bodied moth. It had white wings with black spots, it camouflaged well on trees with light coloured bark and lichen. The black bodied moth was rare.

However, during the industrial revolution, the tree trunks turned black from soot. Now the moths couldn't blend in. They became easy prey for the birds. The moths that had more black spots were harder to see against the blackened trunks, survived, producing darker and darker offspring with each generation, until nearly the entire population of peppered moths became black.

When the industrial pollution lessened and the trees regained their light colour the white-bodied moth returned.

A simple model on the next page may sufficiently explain the concept of evolution through natural selection

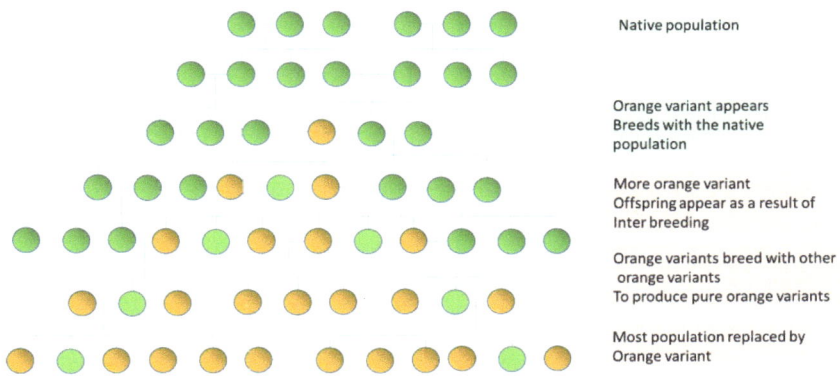

Imagine a population of green coloured insects that live on a tree with orange flowers, an orange coloured offspring is born in the population through a random mutation. The orange variant can camouflage better on the orange flowers and has a survival advantage.

It breeds with the green individuals and produces more orange offspring. Since they can survive better than the green ones they begin to increase in numbers and the variation spreads through the population, until the population is almost entirely replaced by orange individuals.

A green species gives rise to a new orange species.

THE MODERN THEORY OF EVOLUTION

The theory of evolution today is referred to as "Current evolutionary synthesis".

It is fundamentally rooted in the Darwinian framework. Variations appear randomly in offspring of existing species. The offspring with good variations survive and spread the variation to a larger and larger population. Accumulation of over several generations, leads to emergence of new species.

The question is how do offspring from the same parent acquire variations?

A number of new observations have been added to explain how animal acquire genetic variations and start differing from their parents. These could be **Mutations:** random sudden changes occurring in the genome (the DNA) giving rise to a new, unexpected trait or· *Gene migration,* which is the transfer of genetic variation from one population to another. This usually occurs due to the migration of individuals of one population to another and bearing offspring. More migration means more homogeneity between populations. Something that may happen between two similar flocks of birds.

Another process that brings about variations in a population is *sexual selection*: The fittest and the strongest members of a species survive competition within a group of animals and produce offspring that are stronger and fitter, who in turn would have a higher chance of survival and mating. Lastly the phenomenon of *genetic drift,* which is the fluctuation in the frequency of an existing gene variant in a group due to the effect of random sampling. The random sampling usually arises from the fact that whether an individual within a group is able to reproduce successfully is not predictable and there is an element of chance.

The first two processes, mutations and gene-migration create variation in a species, whereas the latter, natural selection and genetic drift govern the distribution of these traits amongst populations.

Although these new mechanisms have been discovered, they only explain the how individuals amongst a species acquire variations, but the fundamental paradigm of the theory has remained unchanged, which is, that the offspring are born with variations, the offspring with advantageous variations survive the struggle for existence. The fittest survive and produce better offspring. Over time these variations accumulate to give rise to new species.

It can be summarised in its definition:

"Evolution is changing in the heritable characteristics of biological populations over successive generations"

At this point, one may delve into two terms used in discussing the evolution

Evolution has been divided into micro (small) evolution and macro (big) evolution to describe the magnitude of change.

Microevolution refers to small variations occurring over smaller timescales and, ***Macroevolution:*** refers to major changes over massive timescales.

However, there is a general agreement that the basic mechanism of macroevolution is the same as microevolution. Macroevolution is the result of the compounding effects of small changes over longer periods of time.

Another concept that emerged from the theory of evolution was that of **common descent**.
As Darwin contemplated on his theory his thoughts began to wander back in time. He thought of ancestors of the species, their ancestors and their ancestors, the fossils of strange animals that were extinct. He concluded that ultimately all living forms that

exist had a common ancestor. (In modern scientific language it is referred to as the Last Universal Common Ancestor or LUCA).

It is a process of accumulation of small changes (variations), that occurs over generations that create major variations over timescales of millions of years.

These changes occur in a random fashion and persist if they incur a survival advantage to the bearer.

So the LUCA was a single-celled organism that over millions of years evolved into a simple multicellular organism, which then gave rise to the creatures with true tissues, the invertebrates which then gave rise to vertebrates through a random process of variations and mutations. It would also mean a gradual, continuous progression from simple to complex life forms with a continuum of intermediates between each transition.

SMOOTH TRANSITIONS OR SUDDEN JUMPS

THE PERCEIVED (AND THE REAL) PROCESS OF EVOLUTION

The concept of evolution appears to be fairly simple. Single-celled life forms gradually evolved into more and more complex forms over several million years through the process of natural selection. The species which failed to adapt to the changing environments became extinct, while those that survived continued to survive and evolve.

There is this picture in our minds: the idea that the prehistoric animals were somewhat unsophisticated, and they gradually evolved into animals of increasing complexity until modern day animals including humans.

The illustration below represents a simplified view of the process

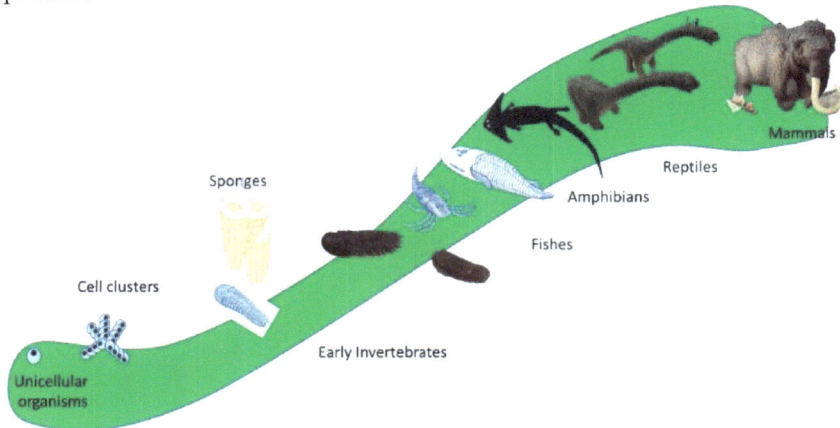

The perceived process of evolution: single cell animals evolving to sponges and worm like animals, followed by more complex invertebrates such as trilobites and sea scorpions. Invertebrates evolving into early and then more complex vertebrates (fishes). Fishes giving rise to amphibians. Amphibians colonising land as reptile. Reptiles then giving rise to birds and subsequently mammals.

However it may surprise some readers that it is not true. Complex animals actually did not emerge in an elegant gradual process as believed. They emerged with a bang. Scientists call it the **Cambrian explosion**. It refers to the period of 485 to 541 million years ago in Earth's history.

No one really knows how this happened. But the story goes, that on the 30th of August 1909, palaeontologist Charles Walcott stumbled upon a remarkable find, The Burgess Shale, a fossil-bearing deposit in the Canadian Rockies of British Columbia.

What he found there shook the scientists all over the world. There were remains of a variety of complex animals that had seemed to appear almost simultaneously around 450 to 500 million years ago.

Not only that they were numerous in count, they represented practically all major animal groups found today, of what scientists refer to as the phyla* (singular: Phylum). There were arthropods (trilobites, sea scorpions) molluscs (nautilus, brachiopods), annelids (worms, and leeches), echinoderms and early vertebrates.

[*A *phylum refers to the major animal group in taxonomical hierarchy that has a body plan distinct from all others. For example arthropods are the animals with exoskeletons and jointed legs (insects, spiders, crabs) that can't be confused with molluscs that have soft bodies (such as snails, octopus and cuttlefish*]

More surprising than the fact that representatives of all phyla of multicellular organisms appeared simultaneously, is the fact that no new phyla have been added to the animal kingdom since.

Here we need to correct our earlier perception of evolution.

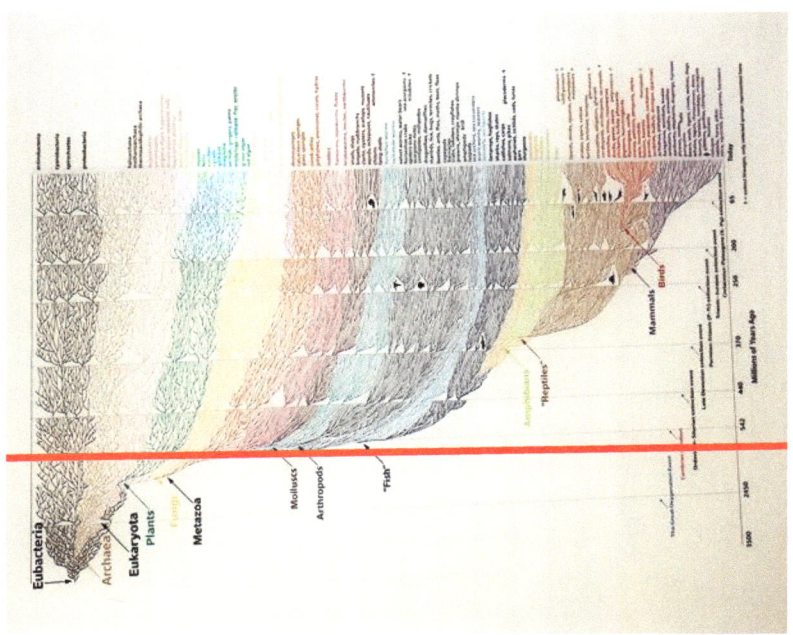

The modern evolutionary history: *Each colour represents individual phylum. Several phyla appearing simultaneously during the Cambrian period (red line). Of note is that no new phyla have appeared since then and evolution of new species strictly stays within each phylum. (source: display at LKC museum of natural history)*

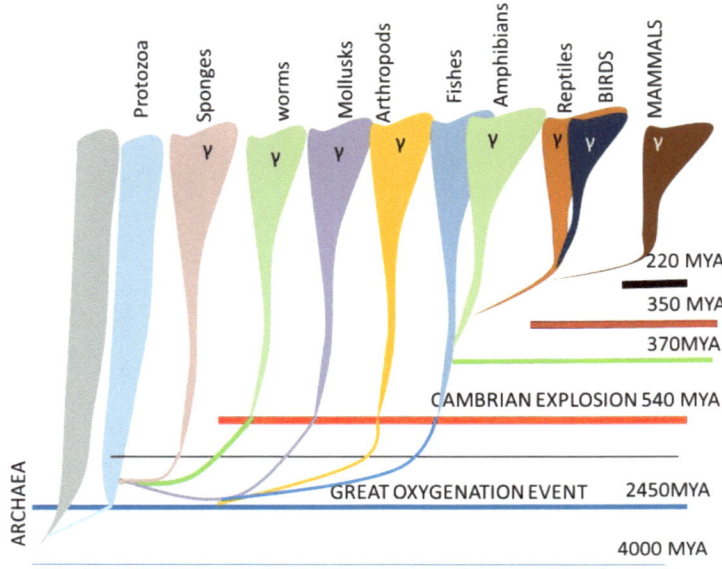

*MYA: million years ago

A simplified version of the previous chart: *evolutionary history. Each colour representing a lineage or a major group of animals (phylum).*

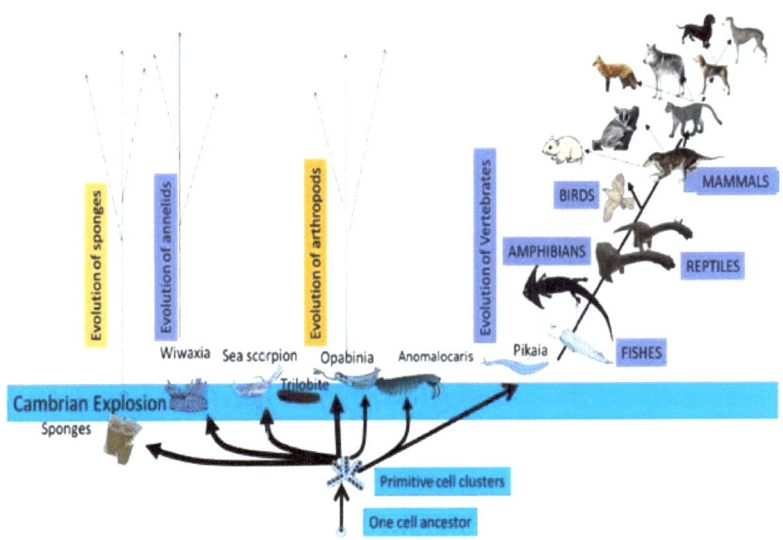

A visual representation of evolution: *appearance of several classes of animal phyla (Sponges, annelids arthropods and vertebrate ancestors shown here for simplicity). This was followed by linear evolution of each phylum*. Only vertebrate evolution is detailed.*

The evolutionary feat included several designs of brains, nervous systems, locomotor systems, mouth parts, limbs, eyes and distinct modes of reproduction.

The single cell organism organised itself into sponges, another branch organised itself into organisms with true tissues representing all major body plans in existence today.

[to be specific there are two major types of body plans : the diploblasts or the bilayer* organisms with a functioning nervous system (like flat worms) and the triploblasts*, with three layered body systems , which again had two distinct embryologic processes and were grouped into two groups protostomes and deuterostomes. Protostomes** are divided into the Ecdysozoa, (e.g. arthropods, nematodes) and the Spiralia, e.g. molluscs, annelids, Platyhelminthes, and rotifers. The deuterostomes** include star fishes, hemichordates and chordates.]

The **Cambrian explosion** is not some obscure secret. It is well acknowledged in the scientific literature and remains an enigma. The sudden and simultaneous appearance of so many categories of animals hardly fits the Darwinian model of gradual evolution.

It is not only a quantum leap in terms of evolution (single celled creatures transforming into fully developed animals) it took place in an extremely short period of time in the history of life on earth.

Since this finding disrupted the idea of the gradual march of evolution in small steps of natural selection, scientists have proposed several hypotheses that would bring this phenomenon back into the evolutionary framework.

All the hypotheses were meant to explain how the evolutionary process could have been accelerated during this time frame and how, the most dramatic event in the history of life was achieved in the shortest period of time.

These are the main theories:

Plate tectonics and greenhouse effect: According to this theory major changes in environmental conditions could have driven evolution. It is well accepted that rapid major environmental changes could cause extinctions (such as the dinosaurs), as they do not allow the species enough time to adapt. It is difficult to accept in the same breath that environmental changes could create new genetic blueprints leading to a rapid emergence of new animals.

Plate tectonic driving evolution in the Cambrian period (Kirschvink, Ripperdan, & Evans, 1997).Eyles & Januszczak, 2004; Kerr, 1993, 1274-1275;

Arms Race between predator and prey: Emergence of predator-prey in the ecosystem triggered diverse adaptations, including shells and new modes of locomotion *(Campbell, Reece & Mitchell, 1999, p.596; Kerr, 1993)*. The problem is in the Cambrian ecosystem, several predators and prey species had appeared simultaneously. The theory may explain their evolution *after* they had appeared but not how the predator and prey had appeared in the first place.

Mass extinction: Mass extinction of the preceding Ediacaran ecosystem, leading to adaptive radiation of surviving species *(Kerr, 2002; Knoll, 2003)*. The concept seems to be contrary to the idea of struggle for existence. A mass extinction of earlier biota would reduce survival pressures, and would discourage adaptation (this theory is in a way, diametrically opposite to the previous theory)

Reaching Complexity threshold: Animals reached a certain level of biological complexity necessary for rapid evolution *(Conway Morris, 1998; Kerr, 1993; Raff, 1999)*. Similar to the above theory, how was the complexity threshold reached in the first place? We also just heard that the evolution was driven by the *disappearance* of the biota of the preceding Ediacaran period.

Emergence of Set aside cells: A proposed initiator for the Cambrian explosion was the emergence of "set aside" cells *(Davidson, Peterson & Cameron, 1995; Hecht, 1995)* that are held in reserve during larval development, but later are activated to build the adult body form. The theory fails to address the reason why these cells appear and then acquire the genetic blueprint to construct an adult body.

Appearance of Hox gene: Hox genes have been found to direct the development of animal body plans, are common across the diverse animal phyla, and so may have been present in the common ancestor of bilaterally symmetrical animals. The emergence and duplication of Hox genes could be the primary explanation of the Cambrian explosion. A ***fortunate mutation*** that created perfect body-plans of approximately 30 different types.*[What we are observing is simply a common gene shared by bilaterally symmetrical animals, we are calling it a mutation to satisfy the paradigm of evolution]*

It is interesting to note that these theories are ignoring the fundamental question: Through what biological process did a unicellular organism assemble itself into a multicellular animal?

How does a single construct a body, wire it with a nervous system and create various other organs, link them together and make an animal, all without a plan?

There is another set of theories that try to cover that gap, which we will discuss in a later chapter.

A beautifully written discussion by Dr. Stephen. E Jones can be found on:

http://creationevolutiondesign.blogspot.com/2006/02/cambrian-evolution-sic-of-animal-body.html

A CLOSER LOOK AT THE WORD "EVOLUTION"

The theory of evolution holds immense power. It has been likened to 'universal acid' that can cut through any argument. Where does such power come from?

If we look at the definition of evolution, there are several of them in scientific literature. Some examples are as follows:

"Evolution is change in the heritable characteristics of biological populations over successive generations".

[Hall & Hallgrimsson 2008, pp. 4–6 "Evolution Resources". National Academies of Sciences, Engineering, and Medicine. 2016. Archived from the original on 2016-06-03.]

"All changes in the characteristics and diversity of life on earth throughout its history"

[page 301 Biodiversity for NUS Mcgraw Hill publications 2017]

"Evolution: broadly defined as any instance of change over time. More specifically in a biological context it is the process of descent with modification that is responsible for the origin, maintenance and diversity of life"

[Carl T Bergstrom, Lee Alan Dugatkin, Evolution second ed WW Norton and company]

The definition is incredibly broad and all inclusive.

His concept becomes apparent if one can look at some examples of evolution.

The transformation of wild dogs into several varieties of dogs satisfies the definition of evolution, as well as the transformation of reptiles into mammals. And so does the transformation of single cell animals into multicellular animals. There is a massive difference of magnitude in these three examples.

Emergence of varieties of dogs from wild dogs

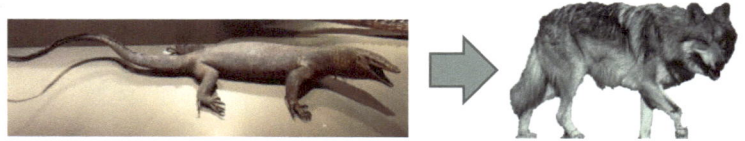

Egg laying reptiles giving rise to child bearing mammals.

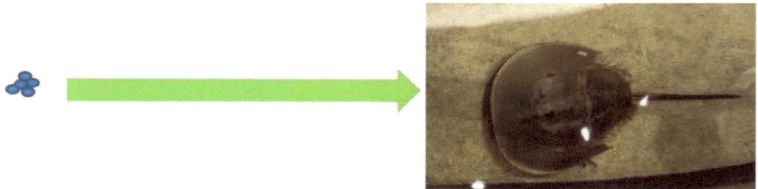

One cell animals giving rise to complex multicellular animals with complex organs and reproductive systems.

Emergence of modern dogs from wild dogs, emergence of mammalians from reptilian ancestors, and origin of multicellular animals such as the horseshoe crab from one cell animals, all are clubbed under the big umbrella of evolution. All these transitions are considered to have occurred through the evolutionary mechanism of natural selection.

In the first case the modern dog essentially has exactly the same internals structure as its wild ancestor. The difference is in the size and minor changes in the shape of the snout or the size of its limbs

However in the latter case, the difference between the two animals is massive. The mammalian physiology of the mammal including the body structure, the internal organs (heart, lungs, brain) and the reproductive system (giving birth to babies instead of laying eggs) is entirely different from its reptilian ancestor.

In the third example, the magnitude of transformation is even more massive. The transformation of one cell animals into a multicellular animal, in this case a horseshoe crab (an animal that has shared the seas with the most ancient animals during the Cambrian era, and hasn't changed to this day). This transformation involves formation of the entire body, including the brain the nervous system, the internal organs and the reproductive system literally from scratch.

These different forms of evolution are classified as '**micro**' and '**macro**' evolution. Although micro and macro-evolution take into account the magnitude of change, they are considered to be products of the same evolutionary mechanism. It is simply that macroevolution occurs as a compounding effect of small micro-evolutionary changes over huge timespans.

This brings us to the most *fundamental assumption* that lies at the heart of the evolutionary theory, it is that:

The process of natural selection could create new organs including organs of reproduction where none existed, through small successive steps

 and

The process of natural selection could also transform existing organs including organs of reproduction, giving rise to structurally novel forms through small successive steps.

*In other words, **all degrees of transformation** were possible through a gradual process of natural selection, irrespective of the magnitude of that change.*

Thus, based on the theory changes in pigment (as in the case of brown pigeons from grey ones), and the transformation internal organs as well as the reproductive organs (as the case of origin of mammals from reptiles) and the formation of entire animals all were possible through the process of natural selection, and ***the evidence for the pigeons could be used to substantiate the latter two transformations.***

This echoes Darwin's concern in his original manuscript

"If it could be demonstrated that any complex organ existed, which could not possibly have been formed by numerous, successive, slight modifications, my theory would absolutely break down. But I can find out no such case"

The statement was never put to test, and was always assumed to be true.

This assumption is so powerful that it literally forms the backbone of the entire idea of evolution.

It has given the evolutionary scientists unlimited power. Once a fossil is discovered (usually a skeleton or part of a skeleton) one had to find its closest relative. Then fill in the gaps with imaginary creatures (***called ghost intermediates***) in various stages of transformation. One simply assumed that the intermediates would have existed, just that they haven't been discovered in the fossil records.

It may be fairly simple to imagine intermediates but little discussion seems to take place in considering the ***biological feasibility*** of these imaginary creatures.

In other words could these imaginary animals have existed in the real world.

The following analogy might help clarify the idea of biological feasibility.

GHOST CARS

Why feasibility of transformations is an important consideration?

Evolution has been likened to making changes in a car while the engine is running.

Imagine you are an archaeologist in a distant future belonging to a highly advanced machine civilisation. While exploring an archaeological site it has discovered two cars from the past. Within an older stratum of earth, a car with a gasoline combustion engine, and within a more recent stratum was a car with an electric motor. From the similarities in the shape of their 'skeletons' it was clear that they were 'related'.

Further excavation revealed no other cars in between those strata. And searches at many other sites had also found similar cars in those strata, with no other cars in between.

The first logical conclusion would be that the old gasoline cars were **replaced** by electric cars. That would mean that the cars had appeared ready-made, implying that someone was making these cars (a car-maker)

However you were taught a different paradigm. The paradigm was that the human civilisation was a myth and that there were no 'car-makers'. The only way to explain the presence of the new car was that the first car had to evolve. It would mean that there should be several generations of cars in between demonstrating gradual transformation from the older petrol car to the electric car.

However the problem remained. No one had discovered such intermediate cars.

In order to fit the findings to the paradigm, the gaps had to be filled with a series of drawings that represented the transformation of the petrol car into the electric car.

In doing so, you choose to ignore a fundamental problem. The intermediate cars (with part of the engine being a combustion engine and part of the engine being an electric motor) would be non-functional, and couldn't have existed in real life. In other words you did not test the **feasibility of the transformation.**

*(We can exclude the Hybrid cars of today which have **two separate engines**, a combustion engine and a separate electric motor and you can switch between the two, an intermediate would be where part of the same engine is a combustion engine and part of it is an electric motor).*

If one were to test the feasibility of the evolutionary transformation and demonstrate that existence of such intermediates was not possible, then one would return to the original conclusion, that the cars were replaced by car makers, and the logical approach would have been directed to looking for the car- makers, i.e. the human civilisation, rather than adhering to the dogma that there were no car-makers and filling in the gaps with hypothetical intermediates.

Similarly in order to prove the process of evolution in our world one would have to line up our findings from various strata and look at the various transformations, and then test them for their biological feasibility.

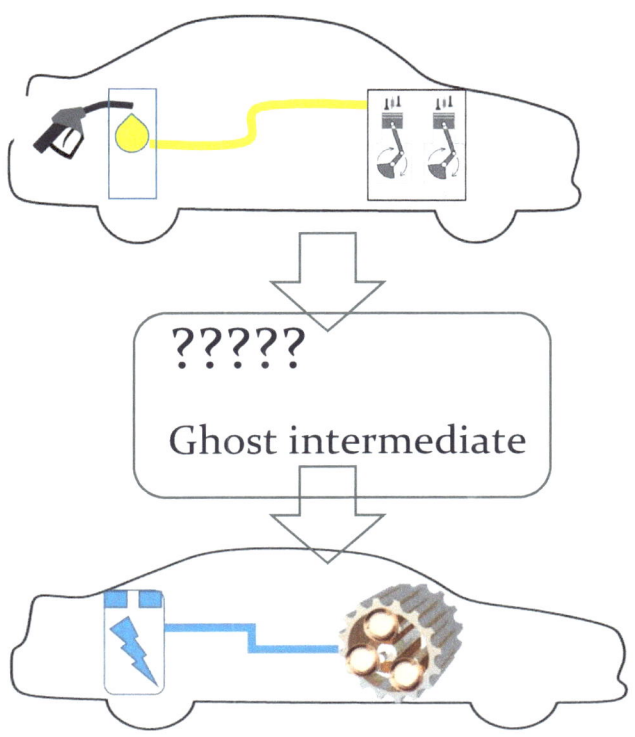

On top is a gasoline car with a gasoline tank gasoline flows to a combustion engine via a pipe and the combustion engine to be ignited. The contains several pistons to convert the ignited gasoline into a spinning motion for the wheels.

Below is an electric car with a battery a cable brings electrical power to motor (containing several copper coils and a magnet. The current flowing in copper coils converts the electricity into a spinning motion for the wheels.

In order to evolve one must fill the gap with and intermediate.

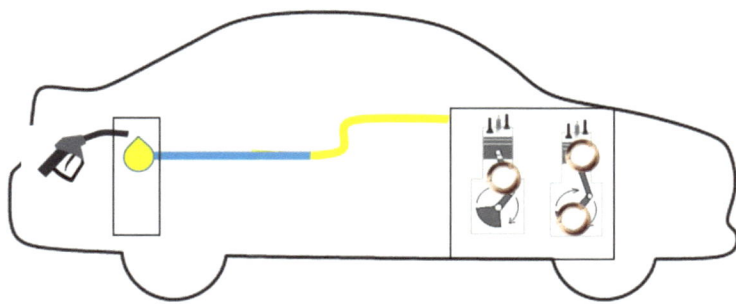

The Ghost Car : *the gasoline pipe is partially replaced by an electric cable, parts of the pistons are randomly replaced by coils, and the magnet has not evolved yet! Such a car would be essentially non-functional, and could not have existed.*

LEVELS OF BIOLOGICAL TRANSFORMATION: A NEW MODEL FOR LOOKING AT EVOLUTION

If we were to take the example of a modern animal, and trace the origin to a unicellular ancestor, we would encounter several levels of transformation which would range from small morphologic transformations to major transformations that involve changes in internal organs.

Take the example of a dog. Assume it is a Husky. From a taxonomic point of view, the dog is housed in a stack of groups and subgroups as follows:

Kingdom	Animal	All animals
Phylum	Chordata	Fish, Amhibians, Reptiles, Birds, Mammals
Class	Mammal	Bats, monkeys, cats, dogs, whales
Order	Carnovora	Cats, Lions, dogs wolves bears
Family	Canidae	Dog, wolf, fox
Genus	Canis	Dog and wolf
species	Familiaris	Dog

A husky is a breed or a subspecies from a bigger group formed by all dog species in the world.

From evolutionary point of view the dogs likely descended from wolves. The wolves form the part of a group called canines along with foxes and bears. Canines differentiated from ancestral carnivores (the *Miacidae*) that were ancestors to both, big cats and wolves.

This big group of carnivores possibly descended from a more general group of primitive mammals along with other groups such as apes, rodents, herbivores etc.

Until now all the transformations have been within the same major group, the mammals. All of the animals till now have possessed the same form of reproduction and the same basic set of internal organs (with some variations) and the same basic physiology.

However based on similarities in the structures of mandible, teeth and skeletal features, mammals are believed to have descended from reptilian ancestors. Here, although the body structure may be similar between early mammals and reptiles in terms of the limbs and the skeleton, they had a different physiology (warm blooded versus cold blooded, body hair versus scales) and they had a distinct forms of reproduction. Reptiles are born out of eggs whereas the mammals develop within the bodies of their mothers and depend on a uterus and a placenta.

Similarly the reptiles which lay shelled eggs possibly arose from amphibious ancestors (like salamanders) that laid soft jelly like eggs.

The amphibians arose as descendants from the fish which lived and could survive only in water.

The fish arose from a remote ancestor for all vertebrates, called the *Pikaia* that inhabited Cambrian seas along with several other animals, the first invertebrates (animals without backbone).

Finally all multicellular animals arose from single cell ancestors.

If one looks at the process of evolution. Several distinct levels of transformation can be seen in the emergence of the dog.

Thus the earliest transformation in the evolution of the dog was the emergence of unicellular organisms or primitive clusters of unicellular organisms to multicellular organisms. This was a massive leap and involved the design of the entire body. The cells had to turn into specialised cells like nerves or muscles, these cells had to organise themselves into tissues like the brain (or early ganglion) or the muscles. The tissues then had to integrate into organs so they can function as units. Then all these organs needed to organise themselves into organ systems that could provide the necessary function for the organism and then all these systems had to be packaged in a body cavity that separated the entire internal system from the external environment and yet be able to extract essentials such as food and nutrition from the environment.

Once the bodies were formed they also needed the most essential component for evolution: the reproductive system. A method of producing cells that would meet and then undergo programmed transformation into a younger version of the parent capable of independent survival.

In this case we take the first transformation as the emergence of the Pikaia (A small eel like ancestor of all vertebrates) from simple cell clusters.

The second transformation would be the transformation of the Pikaia into fishes which then evolved limbs and lungs to become the amphibians. They were dependent on their aquatic environment for breeding.

However if we observe the next step, the ***origin of reptiles*** from the amphibians a major transformation in the reproductive system took place.

The water based reproduction of jelly like eggs couldn't work on land. The transformation required the development of land based reproduction that required production of eggs with shells.

The birds were also descendants of the reptiles. It was the reptiles that developed feathers, but the reproductive system remained similar.

For the next level of transformation of reptiles to ***mammals*** a placental system of reproduction had to evolve. The egg laying system had to change to a system of carrying the embryo within the body inside a uterus and giving birth to live offspring.

The next transformation would have been the transformation of ancestral mammals into *various groups of mammals,* such as rodents, primates, carnivores, ungulates etc. Essentially it was a change in the physical structure with changes in the morphology or the shape of organs, but the physiology and the reproductive system remained same.

The further transformation involved far lesser degrees of change. Carnivorous ancestors gave rise to the wolf and cat like descendants and then wolves to dogs, which through the process of selective breeding gave rise to the husky, the greyhound and the poodle.

We notice that there are several levels of transformation before a unicellular ancestor could give rise to a dog.

We can analyse evolution based on levels of transformation seen above. And label them as **alpha, beta and gamma** for convenience

Alpha (α) Transformation: Transformation of unicellular life forms into animals with multicellular bodies (metazoa).

This transformation involved the organisation of cells to tissues, tissues to organs, organs to organ systems unique to each body plan, and a unique reproductive system for each body plan. This also represented the **origin of all major phyla** of multicellular animals

Beta (β) Transformation (Origin of classes): Emergence of novel physiologic forms (categorised as 'Classes') from existing forms within the phylum. This change involved the transformation of the reproductive system [amphibian (eggs without shells) to reptilian (shelled eggs) and reptilian to mammalian (live birth) transformation within the chordates] and modification of organs (reptiles have kidneys that look and function differently from mammals, similarly the structure of the heart and brain is different in both groups)

Gamma(γ) Transformation (Origin of Orders): Emergence of novel morphologic forms within the class. This involved changes mainly in body or organ size or proportions, **without transformation of organs or reproductive system.** (For example emergence of widely differing groups within mammals from a common mammalian ancestor, such as rodents, primates, ungulates, carnivores etc. or a wide variety of reptiles from a primitive reptilian ancestor)

Gamma2 (γ2) Transformation (Origin of families): subset of gamma transformation, with emergence of Families within the order (separation of cat family from the dog family)

Gamma3(γ3) Transformation Origin of Species: Emergence of genera, or groups of species within the families (separation of wolves, foxes and dogs and dogs giving rise to new breeds)

Gamma γ: Changes in external features only dog species arising from wild dogs.

Beta β: Changes in internal organs and reproductive systems (egReptiles giving rise to mammals)

Alpha α: Formation of entire system of organs (One cell animals giving rise to complex multicellular animals)

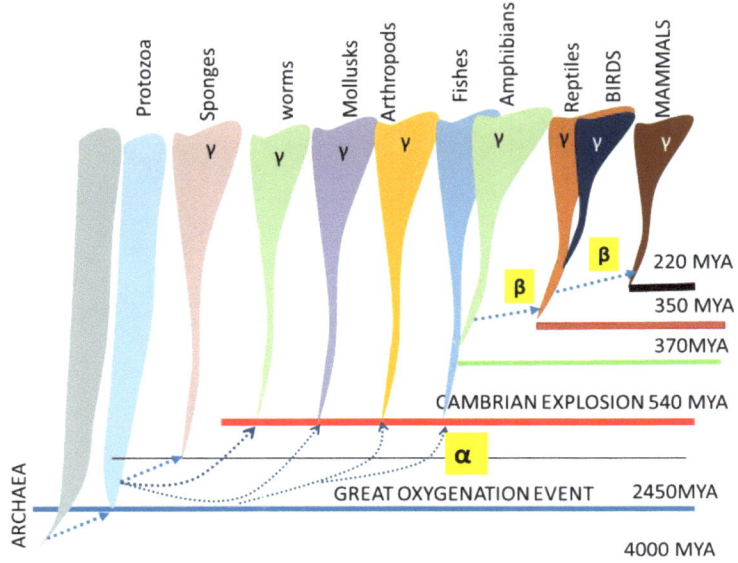

Major transformations: *Alpha(α) transformation: single cell animals to multicellular animals. Beta (β)transformations: transformation of reproductive systems (with the origin of reptilian and mammalian reproductive systems form amphibian and reptilian systems respectively).Gamma (γ) transformations within the same class of animals.*

(*MYA: million years ago)

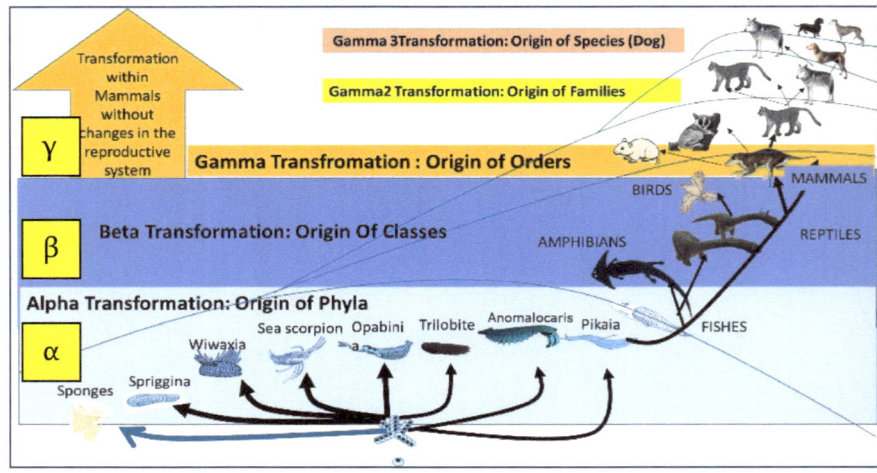

Levels of transformation involved in the origin of dogs *Origin of primitive vertebrates from single cell animals (α transformation), origin of reptiles from amphibians (β) transformation, origin of mammals from reptiles (β) transformation, origin of various orders within mammals (γ) transformation. Origin of carnivore family (γ2)Origin of dog species(γ3)*

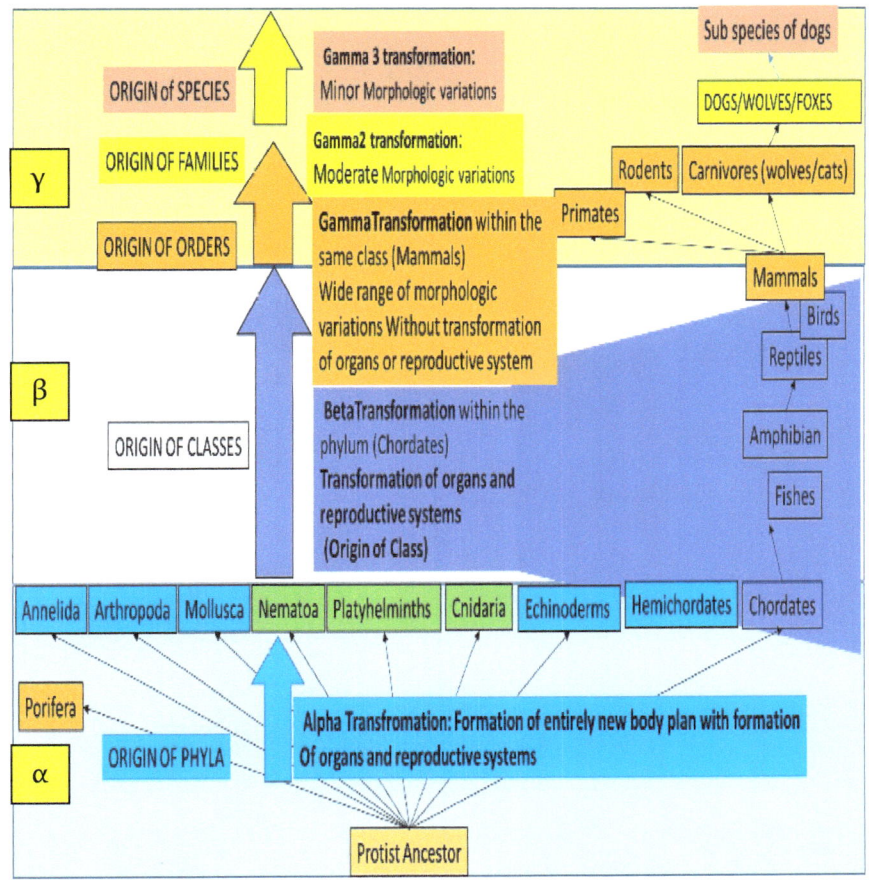

Schematic representation of levels of transformation in the origin of dogs

Table summarising levels of transformation seen in evolution

Level	Evolutionary step	Biologic basis of transformation
Gamma 3 ⇧	Origin of species	**Small phenotypic transformation (wolves/foxes/dogs)**
Gamma 2 ⇧	Origin of families	**Moderate phenotypic transformation (cat vs dog family)**
Gamma Transformation ⇧	Origin of orders	**Major phenotypic transformations *within mammals* without transformation of reproductive system** (primates/bats/rodents
Beta Transformation ⇧	Origin of classes	**Transformation of reproductive system and modification of organs (eg. reptiles to mammals)**
Alpha Transformation	Origin of Phyla Fist metazoan from unicellular organisms	**Origin of new body plans from unicellular cells including organs and reproductive system**

THE CONDITIONS FOR EVOLUTION BY NATURAL SELECTION

The stratification seems to be just another way of representing the evolutionary tree. However this helps us to examine **whether each of these levels of transformation was biologically feasible** through the process of Darwinian evolution.

As we know, evolution is defined as "...change in the heritable characteristics of biological populations over successive generations."

Based on the definition of evolution we can infer that for an animal to evolve

- There needs to be a population of those animals.
- The members of the population should be able to survive in their immediate environment and be able to breed for several generations.

Hence evolution between one form to another would only be biologically feasible if

1. The first, and all subsequent generations were born alive, and were able to survive in their environment (viability)
2. Each generation was able to reproduce successfully. (fertility)

If any of the conditions were violated, (failure of survival or failure of reproduction) the species would become extinct and fail to evolve further.

We take survival and reproduction for granted. However it requires several conditions to be fulfilled before these conditions can be satisfied.

In order to understand the complexity these concepts (viability, survival and reproduction) we need to take a closer look at the living animal.

WHAT DOES IT TAKE FOR AN ANIMAL TO BE VIABLE? JUST PERFECT FORM, AND FUNCTION

Evolutionary theory speaks about the common origin of all animal and plant life on earth.

A multicellular organism today is an offspring of its parents, who are the offspring of their parents. We can trace it all the way back to the very first animals representing the phylum somewhere in the Cambrian period. We have labelled this population the *pioneer population*. However far in the past, irrespective of the era, in order to prove the theory of evolution it would be imperative to demonstrate the biological feasibility of the very first transformation.

The transformation from single cell to a complete multicellular organism (the α- transformation).

These pioneer populations had to be ready to survive from the point of emergence, and reproduce in order to set the process of evolution in motion.

However survival is an extremely complex issue. It depends on a delicate balance between body systems and environment. **Can a randomly designed animal survive?**

Imagine a population of gold fish in a pond. They are a part of the pond ecosystem. They depend on the oxygen produced by the plants and feed on certain small worms that are present in the pond. There are microscopic creatures in the water as well as larger crustaceans such as crabs and turtles in the pond.

One can observe the relationship of the fish to its environment through some simple experiments.

If we move this population of goldfish into a tank of sterile water with no plants or a source of oxygen, soon all the fish would begin die for lack of oxygen. This is because they are born with a respiratory system which can function only within a certain range of oxygen levels in the water. Once the entire population dies, it won't propagate to the next generation or evolve into a fish that can survive without oxygen.

Similarly, if we remove all the worms from the water leaving only the microscopic organisms and the larger animals such as crabs and the turtles, the population will starve to death. The reason is that its mouth parts are of certain size and capacity, and can only feed within a narrow spectrum of food that is available. The entire population would perish without any further propagation or a chance to evolve. Again this illustrates that an organism must be endowed with mouth parts suited for the nutrition already available in the ecosystem.

The mouthparts of the fish can only be used to feed on food particles of a certain range of sizes. This has a major implication when considering the emergence of pioneer population in an ecosystem. ***The body parts must be in synchrony to the type of nutrition available in the environment***

Now imagine we transfer the goldfish to sea water. The fish population would perish in a short time. Again, that is because its organs are structured to process fresh water. They fail in salt water.

We can infer that it is not possible for the fish to adapt its essential functions within a single lifetime. It is only when this population is able to survive for several generations, and has a gradual exposure to environmental changes, that it can evolve to exploit a new environment.

In the second set of experiments we can make changes in the fish itself.

We collect fish's eggs and open them before they are ready to hatch. Even if it is in its natural environment with abundance of food and oxygen, the embryonic fish simply die, simply because they are yet to form all the organs essential for survival. The premature embryo just doesn't have the capacity to survive even in the most ideal environment. The survival of the immature embryo is dependent on the environment and nutrition provided by the egg.

Now suppose we perform a genetic experiment and mutate the fish to have a small tail and fins. It would be unable to swim, to find food or escape predators and eventually die.

Similarly if we create a mutant fish with a deficiency in the brain or the nerves that control the muscles in the tail, the fish will be born with a paralysed body, resulting in starvation and death.

We could keep everything the same and we make its liver larger and heavier, again it would compromise it buoyancy and the ability to swim and find food leading to death. On the other hand If we create a fish with a rudimentary liver, it would be unable to survive beyond birth.

Thus we can conclude that when an organism is born, in order to survive its immediate environment, it needs to be equipped all the necessary organs in perfect form and function in relation to the entire body plan and in relation to the environment.

These considerations don't matter now. Several fishes may be born with deficiencies today and may have perished, however it would matter significantly when the first fish was born on Earth. In fact it is true not only for fish but for any other body-plan that came into being in the primitive Cambrian oceans.

One would be quick to apply the concept of natural selection, and argue that the fish may be born imperfect and over several generations the fish would develop an ideal form.

One could state the example of the Jawless fish, which then appear to have evolved to develop jaws.

However the fundamental flaw in the argument. ***The Jawless fish was not an underdeveloped form of the jawed fish***. It was fish that was perfectly designed for feeding without jaws. It had a mouth which was perfect for sucking in water and filtering food particles, and soft worms that didn't need breaking or chewing. It had a perfect body to swim, breath, and find its food. It was completely adapted for survival in its ecosystem, to the water, to the food available. It is evident by the fact that many of the now extinct jawless fish inhabited the prehistoric oceans for several million years.

This is not to say that an animal can't adapt and change in response to changing environment.

It is to say that that in order to evolve, it has to be, born fit, and equipped with necessary organs for survival and breeding in its primary environment; only then can it evolve to adapt to a new environment, but it ***cannot be born unfit to survive*** in its primary environment ***and evolve to become fit to survive***, simply because an unfit animal would perish without being able to produce any offspring and fail the process of natural selection.

There is another way to illustrate this idea.

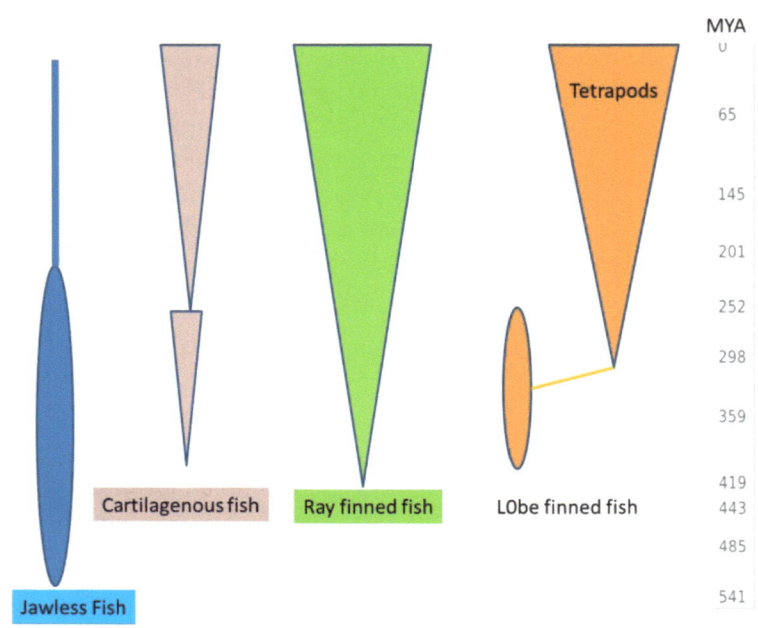

Jawless fish emerged during Cambrian period and thrived for almost three million years, thus could not be considered a group that was unfit for survival

Can one rig up a diving gear while drowning?

Origin of animals in water in several stages could be compared to building a diving gear while being underwater

A person dives into water, and starts drowning. He has magical powers to create mechanical components but he can create one component every hour. He creates an oxygen cylinder. He may have the oxygen, but doesn't have the components to deliver the oxygen to his mouth or the nose. Within five minutes the person is dead.

It is understood that in order to survive the diver should have a complete diving suit with all its components, functioning and connected in the correct order and form.

It illustrates two points, first, an isolated component of an organ or an organ system does not support life, and second the entire life support system must be present before the animal emerges in its environment.

An animal cannot evolve organs and systems critical to survival after its birth, it has to be equipped with it at the time of birth.

THE MAKING OF A LIVING ANIMAL

The car again

If one looks at the car, it has a certain design, we can call it a body plan. It has a shell that houses the engine. It has a gas tank. The gas tank has an opening made for liquid fuel. It has pipes that take the liquid fuel to the engine. The engine is designed to produce energy from that specific fuel. The capacity of the engine is in proportion to the size and weight of the car. It has a chassis that is again designed for the weight of the car. It has wheels that are circular and are designed to run on land.

It also has a system of battery and cables that are connected correctly to the electrical system of lights and the motors for the windows.

One can imagine an autonomous car, which has vision from cameras that works for certain amount of ambient light. The light is converted to electrical signals, that are 'read' by the on-board computer, which has a program (behaviour) to interpret the signals and control the engine for speed, change direction of the wheels or stop the car.

All the elements are housed elegantly within the body of the car.

In order to be functional (alive), all the elements have to be in place simultaneously in the correct proportion to each other.

If the engine was too small for the weight of the car, it would fail to run. If the computer was partially constructed, the car would crash on its first trip. The chassis too weak, the car would collapse.

Suppose instead of gasoline only diesel was available, and one filled the gasoline tank with diesel, the engine would fail.

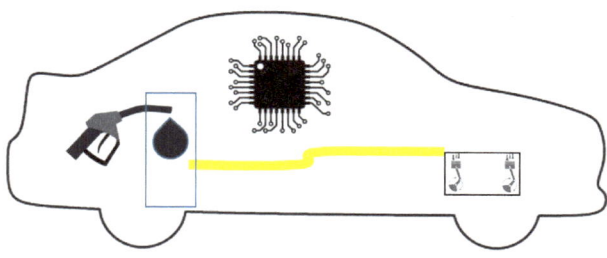

A car born through a random assembly. It has a gasoline engine, the engine is too small for the body and, Diesel is the only fuel available. The computer is partially formed and not yet connected to the rest of the machinery. Such a car would be non-functional; similarly an animal born with imperfect organs and systems, those not precisely adapted to the immediate environment would be non-viable.

The body plan of a living organism is not much different.

An animal can be thought of as a body that encloses several organs. The body maintains an internal environment that protects the organs whereas the organs are responsible for performing functions that make the animal a living entity. The lungs that extract oxygen from air and the heart that supplies blood and oxygen to various parts and kidneys that excrete wastes, and the brain that controls behaviour and movement.

Organs are not just soft slimy lumps of flesh. They are highly complex structures that that are built on several levels of organisation.

The organs are made of several tissues that are structurally and functionally integrated together. The tissues in turn are composed of specialised cells, each performing a distinct function, and the cells themselves are composed of organic molecules that are function in an equally complex manner to make the cell a living entity.

Animals (including their body and the organs) are formed through the extremely complex process of embryogenesis. After the egg is fertilised, it starts dividing. At first it is a ball of cells but then it begins to take shape.

The cells simultaneously divide, change character and migrate in a perfectly co-ordinated manner to gradually take the shape of the animal. Some cells form the skin, while others turn into nerve cells that form the brain, the spinal cord and the individual nerves, and the eyes. Some cells start forming muscles and bones, others start forming he heart, blood vessels and the digestive system, the liver and kidneys.

While so many systems are being formed, they must integrate together. The brain, the liver, the muscle, all the organs must have a network of blood vessels integrated within the tissues, the muscles must have nerves connecting at the cellular level, and they should be attached at correct points to the bones. The brain must have the correct circuits to control the muscles and the correct behaviour to seek shelter and evade predators.

At one point during development, all the organs begin to function synchronously and the embryo becomes 'viable' or capable of a life in the environment without the support of the egg or the mother. At this point it hatches or is born and commences an independent life.

It is a tall order, but it all happens automatically through and extremely complex process of programmed assembly where cells are guided and transformed by the genetic blueprint.

https://youtu.be/3mCgHK-X6lE (Watch a video of embryological development)

The genetic blueprint for embryogenesis comes from the parent's genome and it results in formation of offspring which resembles the parents barring some variations. It is well known that if this process goes wrong the offspring is usually malformed and mostly incapable of survival .It is also well known that an embryo, born prematurely is incapable of survival.

The important point here is that offspring born with immature or partially formed organs are incapable of independent survival, and giving rise to offspring. Once a generation fails it would mean extinction of the lineage.

MAKING OF THE REPRODUCTIVE SYSTEM

The second *quality that is key to the evolutionary theory is the ability to reproduce.*

The reproductive system in the animals is highly complex. It requires the presence of tissues that produce ova and sperms. These are not same as other cells. They are formed through a special process of cell division called meiosis so that they have half the chromosomes as normal cells.

These cells need specialised organs for nursing them and releasing them at the point of fertilisation.

The germ cells need to fuse into a zygote and then start dividing in a highly ordered fashion until a foetus is formed and ready to be born.

During development, barring some sponges, the embryos are nurtured within eggs or in case of mammals, uterus. A system that supports the embryo through the entire process of development until all organs have formed and have matured enough for the foetus to survive independently in the external environment.

The fertilisation as well as the development of the embryo requires highly specific environment. The egg for example contains just the exact amount protein and fat that supplies nutrition to the growing embryo till the time of birth. In hard shelled animals the egg has to be of the correct size for the embryo to grow and porous enough to let air in and prevent bacteria from getting in.

In mammals the process involves the growth of a placenta which is a network of blood vessels that establishes a communication with the maternal blood in the uterus to extract nutrition and oxygen. Simultaneous hormonal changes in the mother allow expansion of body tissues and initiate childbirth precisely when the foetus is ready to be born.

Any insufficiency or abnormality in the process would result in failure of reproduction, and would have a significant impact on evolution. Failure to reproduce would translate to extinction of the species.

If one considers evolution of the reproductive system from a unicellular organism, would the partially developed reproductive system be functional and give rise to the next generation? Unlikely. Similarly if one considers the transition of one reproductive system, from reptile to mammalian, would such an intermediate system be actually possible?

In fact if one considers it carefully, evolution of the reproductive system is a contradiction of terms. It is not possible for the reproductive system to develop through the process of Darwinian evolution simply because evolution itself depends on the ability of the species to be able to reproduce. A partially formed reproductive system would mean failure of the species to propagate essentially leading to extinction.

This would be an important consideration when we are using evolution to explain the origin of reproductive systems in the very first population of multicellular organisms or when we are considering the evolution of the reproductive system from egg-laying reptiles to child bearing mammals.

Thus based on this discussion we can lay down two absolute conditions for evolution.

THE ABSOLUTE CONDITIONS FOR EVOLUTION: THE BIO-FEASIBILITY MODEL

We have classified evolutionary transformation into 5 categories. To be proven true each category of evolutionary transformation should satisfy two conditions of biological feasibility.

The two conditions are:

Viability : Every generation (including the very first) must be born with a fully functional set of organs suited for survival in their environment. Intermediates incapable of survival would perish and fail to evolve further and,

Fertility: For every generation (including the very first) the reproductive system should be complete in all aspects, including the embryologic process. Animals with partially developed reproductive system would be infertile and would become extinct within the same generation. Evolutionary progression to the next stage would fail.

The figure below shows the hypothetical evolution of a multicellular animal from a cell cluster through the process of natural selection and development of organs in stages as predicted by Darwin's theory. The cluster evolves in steps as it acquires a nervous system, reproductive system and various basic organs. However a careful scrutiny reveals that the intermediates with undeveloped organs will neither be viable, nor be able to reproduce. In such situation the very first offspring will abort and fail to continue the lineage.

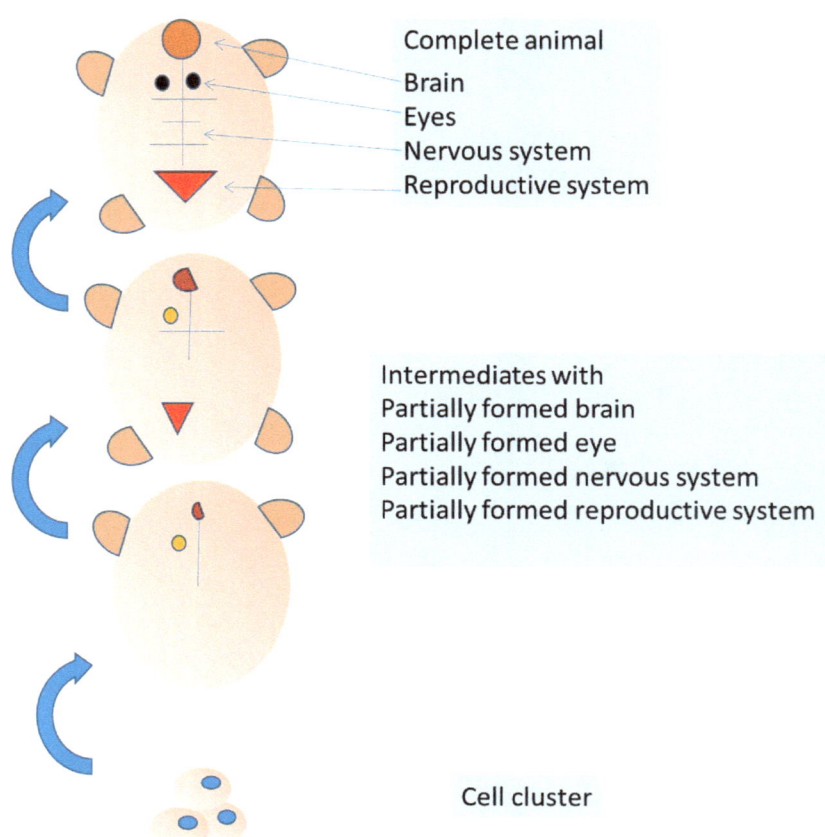

Imagined sequence of development of a multicellular animal based on the evolutionary paradigm. *The cell clusters acquire organs and organ-systems in stages over several generations to create a complete animal.*

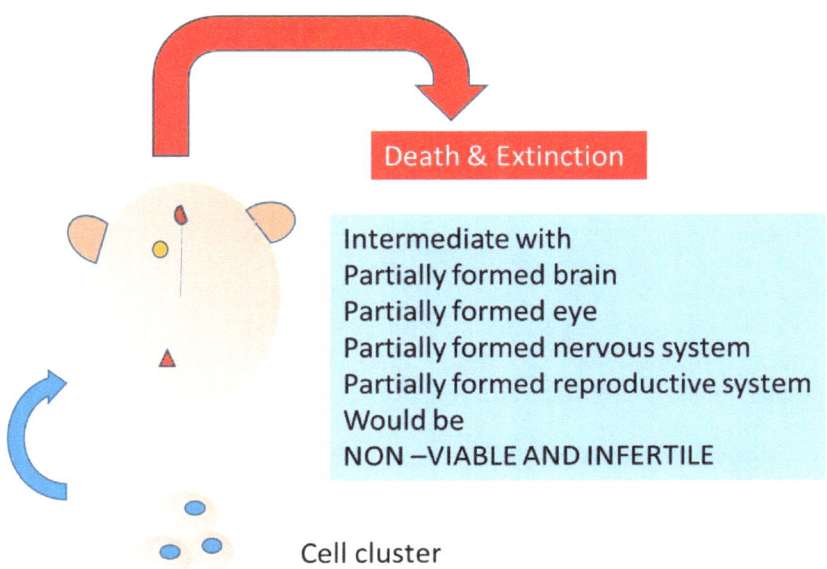

Actual fate of intermediates: Dead intermediates do not evolve

The intermediate with partially formed organs would be non-viable and infertile resulting in death and extinction of the lineage without further offspring.

It is not possible to produce a viable offspring through production of nonviable offspring

The process will repeat itself infinitely until the first perfect animal (viable and fertile offspring) is produced in a single step.

TESTING BIOLOGICAL FEASIBILITY OF THE THREE LEVELS OF TRANSFORMATIONS

The levels of biological transformation mentioned earlier, were:

Alpha transformation: Emergence of multicellular animals from one celled animals.

Beta transformation: Emergence of novel physiologic forms within the phylum involving transformation of the reproductive system (amphibian to reptilian/reptilian to mammalian transformation within the chordates) and modification of organs.

Gamma Transformation: Emergence of novel morphologic forms within the class (body or organ size or proportions) without transformation of organs or reproductive system. (For example evolution of mammals of varying sizes and morphology such as birds from ancient birds, rodents and primates from ancestral mammals)

Gamma 2: Emergence of families (such as the canines and the felines from common ancestors)

Gamma 3 : Emergence of genera, species and subspecies and varieties from an existing species (species of dogs giving rise to new breeds)

We can apply the conditions of viability and fertility to each transformation and see if the transformation is biologically feasible.

THE BIOLOGICAL FEASIBILITY OF ALPHA TRANSFORMATION (EMERGENCE OF MULTICELLULAR ANIMALS FROM ONE CELLED ANIMALS)

The flat worm or the planarian is one of the 'simplest' living creatures. It was the first creature to show cephalization, that is, its nervous system is concentrated in its head end, it has a sense of vision and moves head first. If you watch the planarian moving, it is poetry in motion. It glides on surfaces on thousands of cilia beating in perfect co-ordination. It's not moving in a random, unintelligent manner. The motion is smooth and purposeful. If one were to forget the scale it may have been an anaconda moving through water hunting for food.

The video links below would help appreciate the point

https://youtu.be/w0QzSYQGsnA

https://www.youtube.com/watch?v=nQbRd9kWIEM

The point here is that scientists label the planarian as having a 'primitive' nervous system. It is an unfortunate assumption. It is an extraordinarily complex mechanism. Each nerve cell and muscle cell has to be in the perfect place and orientation. More than that it has a 'brain' that dictates behaviour as it senses and responds through its eyes, touch and chemical sensors. Each movement is dictated by an array of highly complex molecular messengers that carry signals back and forth from the brain.

It has a protrusive pharynx on it belly that it voluntarily controls to suck food and highly specialised 'flame cells' that concentrate toxic molecules and excrete them from the body. And this is one of the simplest multicellular organisms.

In order to be born as a multicellular animal, such as the planarian, the single cell embryo must transform itself into a cluster and then the cells must transform themselves into nerves, muscles, digestive cells and excretory cells, These then must align themselves into organs of correct size and then the organs must make connections to each other, and all that to fit within a definite size and shape of the body of the animal. Some cells would form specialised areas to produce germ cells and reproductive organs.

Evolutionary thinking would predict that a random mass of cells would gradually transform into a more organised mass and over several million years , and ultimately form a multicellular organism.

What does science says about the transformation of one celled organisms into multicellular animals?

Evolutionary scientists believe that unicellular animals: the Choanoflagellates were the ancestors of all multicellular organisms. The Choanoflagellates are single cell animals that have a round collar and a tail (flagellum). These organisms have been shown to form sheet like colonies. They also happen to look like the collared cells seen in sponges.

Choanoflagellate   Choanocyte of a sponge

The reasons to suggest that Choanoflagellates are the ancestors for all multicellular organisms are as follows:

Theory 1:

Choanoflagellates (single cell organisms with a tail like flagellum) form clusters.

Collar cells of sponges resemble Choanoflagellates (they also share common genes)

Hence sponges descended from Choanoflagellates.

Theory 2

Unicellular Choanoflagellates have a tail.

The sperms of multicellular organism have tails.

Hence multicellular animals are descendants of Choanoflagellates

Theory 3

Choanoflagellates can form clusters

We can hypothesize (imagine) unicellular organisms 500 million years ago could have formed form similar clusters

Imaginary cell clusters resemble early forms of embryos of multicellular organisms

Hence multicellular animals are descendants of unicellular organisms.

There are several (six) hypotheses to explain this transformation to fit the evolutionary framework,

The most commonly accepted theory is the *colonial theory*, in which ciliated organisms begin to cluster and form embryo like cell clusters

"The ciliated cells form a blastula like developmental stage of a hypothetical organism (seen in the figure below). In scenario (A) the blastula like organism does not grow because its cells retain the locomotory cilia and thus cannot divide by mitosis.

In the probable scenarios B, C&D, cellular ingression allows growth by internalizing non flagellated mitotically active cells forming three hypothetical metazoans. Three forms of ingression are multipolar ingression (B), unipolar ingression(C) and invagination(D). *(Ruppert, Fox and Bernes Invertebrate zoology 7th ed Brooks/Cole Cengage learning)*

Through this process a hypothetical transition to a sponge like organism may be possible; however true organogenesis and formation of a multicellular animal is unlikely.

The hypothetical (imaginary) cell mass may resemble the early embryo of the multicellular animal, but where does it go from here? There are numerous steps to formation of an entire animal (we may take the example of the 'simple' flat worm).

First of all this cluster of cells does not have the genetic blueprint of a worm, so even if some cells 'accidentally' transform into nerve cells or muscle cells it still remains a jumbled mass of tissue somewhat like an aborted embryo, without the protection of an egg or a uterus. This embryo would neither be able to survive nor reproduce it will simply follow the road to extinction. This process could repeat endlessly, until by some miracle all necessary organs and systems appear in their correct forms in a single step, without a pre-existing blueprint

The biologic feasibility of alpha transformation appears to be unlikely.

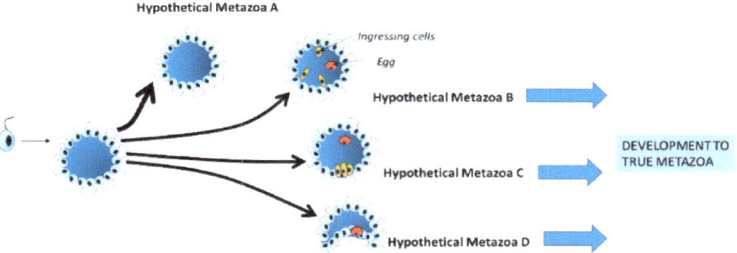

The hypothetical transformation of protozoa into metazoan : Single celled Choanoflagellate forms cell clusters. A. The blastula like cluster does not grow because its cells retain the cilia and thus cannot divide by mitosis. B,C,&D: Three forms of ingression would be multipolar ingression, unipolar ingression and invagination.

Growth by internalizing non flagellated mitotically active cells in three simple hypothetical metazoans

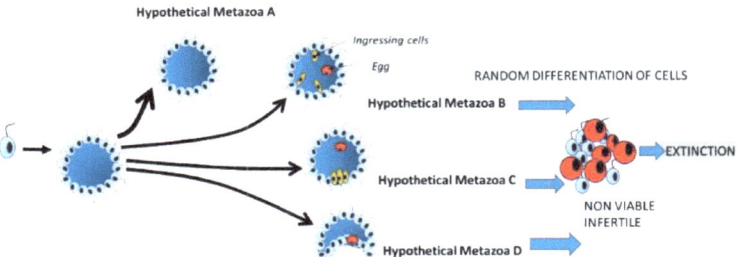

Further extrapolation of the hypothesis: *The three hypothetical forms look like early embryos; however they lack the capacity to reproduce. The cells will undergo an unguided transformation resulting into a random mass of cells which is neither capable of independent life nor capable of reproduction.*

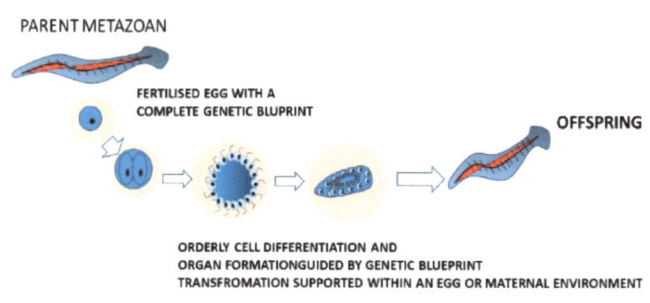

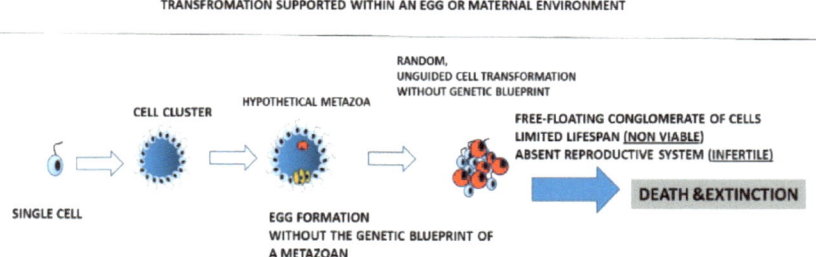

There are some major differences between the embryonic development and evolutionary development. In embryonic development the immature embryo is supported within an egg, and contains a complete genetic blue print for orderly development. In the evolutionary process, the immature cell cluster is not supported by an egg, and does not have a genetic blueprint of a multicellular parent. The development would be random and unguided.

BETA TRANSFORMATION: EVOLUTION OF CLASSES (MAMMALS FROM REPTILES)

We often ask the question: what came first: the egg or the chicken?

If we analyse it from the evolutionary point of view, and trace the ancestry of the chicken, we may find that the first bird may have hatched from the egg of a pre-bird or an almost bird like descendant of the reptiles.

However if I reframe the question: what came first: the ***mammal or the uterus,*** the question becomes more complex.

Fossil 'evidence' based on skeletal similarities indicates that mammals arose from reptilian ancestors, which implies that at some point the reproductive mechanism that was producing hard shelled eggs transformed into a placental system, with a uterus and an entire hormonal support that is required to sustain mammalian pregnancy.

Is it possible to have an intermediate reproductive system that is partially designed to produce eggs and has a partially formed uterus?

or

Could the first placental mammal be born out of an egg? Or could there be a hybrid system which is partly amniotic and placental systems? A foetus partly covered by an eggshell and partly attached via a placenta?

There is a major difference between the egg and the placenta.

The egg is a self-contained structure that does not require a support from the mothers body even when, in some reptiles eggs hatch within the mother.

For the mammal the placenta is literally the lifeline for the mammalian foetus. It contains a dense network of blood vessels that arise directly from the major blood vessels of the foetus. It penetrates the thick wall of the maternal uterus, and creates a root like structure that is bathed in maternal blood. It extracts oxygen and nutrition and releases carbon dioxide, and wastes into the mothers blood, all while its lungs, gut and kidneys are still developing. It is well known that insufficiencies in the placenta jeopardise the survival of the foetus.

If we imagine a transition from an egg to a placental, it would involve gradual disappearance of the egg shell and a gradual development of the placenta. We could imagine an intermediate with the embryo partially covered in a shell and growing a small disorganised version of the placenta.

In a situation of an early or a rudimentary placenta the foetus wouldn't be able to extract nutrition or oxygen that are essential for survival, nor will it have the protection of the egg.

Similarly in terms of physiology the transitional creature would simultaneously possess scales as well as hair, a partial state between a cold and warm blooded animal.

A transitional kidney that is partly reptilian and partly mammalian?

Based on the physiology or the reproductive system, could such transitional forms be biologically feasible.

It is simple to place monotremes (the egg laying mammals like platypus that have no placenta) as intermediates, but it must be noted that these were separate branches in the lineage of mammals. Placental mammals did not descend from the monotremes. A more important consideration is that the platypus may possess mammalian characteristics, but the entire reproductive system is that of a bird or a reptile. The reproductive system itself is not a hybrid of placental and egg-laying systems.

The marsupials on the other hand may have a small placenta compared to other mammals, but it is enough to sustain the embryo till viability, and ready to survive outside the mothers body. If it is less developed or in a rudimentary stage the embryo will be unable to develop at all.

Similarly the transformation of amphibians to reptiles.

The transition from amphibian soft eggs to the reptilian eggs. The amphibian eggs, like those of a frog are soft and jelly like without a hard shell. They are laid in water. The larva that hatches from these eggs is similar to a fish, without limbs and has gills to breath. Over days it develops limbs and functioning lungs. Only then it ventures on land. Neither the eggs, nor the larva can survive on dry land. The reptile's eggs are hard shelled and the embryo within the egg grows and develops until it is ready to survive on dry land, and then it hatches. These eggs are designed for land. Even the sea turtles that spend their entire lives in water need to venture on to dry land to lay eggs.

It is not possible to conceive a transitional form of reproduction where the eggs are partly covered in a shell or the larval form with its gills and some rudimentary lungs that can survive on dry land. It would require a quantum leap to transition from soft eggs to hard shelled eggs.

Based on conditions of biological feasibility, we can conclude that these transformations were not possible. These assumptions may seem conjectural, but we do have a time machine: a periscope to look back in time.

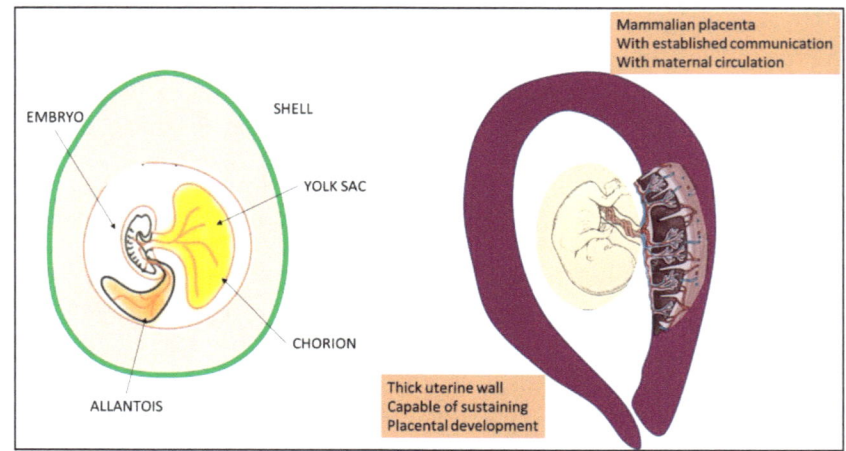

A reptilian egg and a mammalian foetus with a well developed placenta

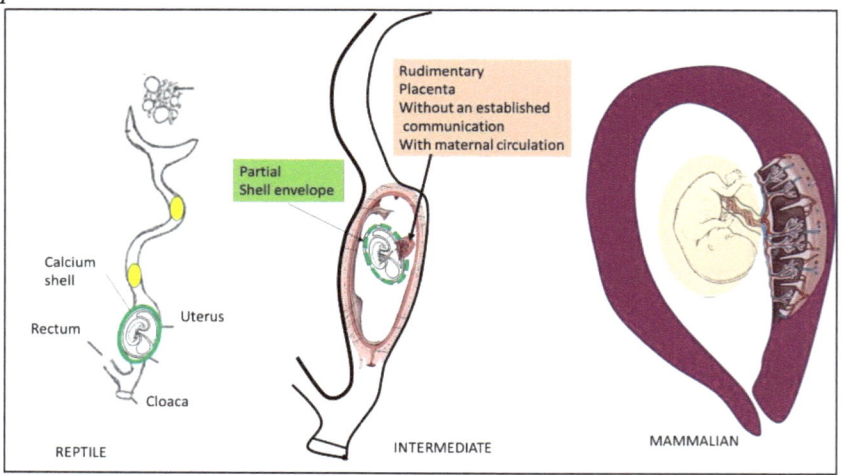

The hypothetical transition between reptilian reproductive systems to the mammalian. The intermediate with a partial shell envelope and a rudimentary placenta, both insufficient to support the foetus

THE 250 MILLION YEAR PERISCOPE

The preceding discussion revealed that the levels that require transformation of organs (at tissue level, not in terms of size and shape resulting in change in the fundamental physiology of the animal) or the transformation of the reproductive system are not possible through the staged process of evolution, however, it is possible for evolution to occur within the physiologic and reproductive boundaries (Gamma transformation).

We have animals today whose ancestry has been traced for 250 million years. This was the era of dinosaurs.

250 million years is a fairly significant time frame since the very first multicellular animals arrived just about 540 million years ago?

Sea turtles have existed for 250 million years, yet to this day they are mal-adapted to their environment. They can't breathe under water, nor can they reproduce .They need to return to land in order to lay eggs. The female turtle is clumsy and vulnerable on land. The eggs, the hatchlings get devoured by land animals and birds, until just a few offspring make it back to the ocean.

These are two major threats to survival and yet **there has been no evolution of the respiratory organs or the reproductive organs**

The turtle had a relative, the Ichthyosaurus, a dolphin like reptile. Fossil records suggest that it was viviparous, capable of live birth in water.

Even with the possibility of live birth within reptilians, the turtle has been unable to evolve its reproductive system.

A parallel example can be cited for mammals that have evolved to inhabit the oceans, the whales.

Whales have inhabited the oceans for 50 million years. Their body has evolved to fishlike shape for efficient locomotion, yet modern whales have not been able to evolve an underwater breathing mechanism like fishes. They retain the mammalian system of lungs. Like turtles they need to surface every few hours to survive.

The whales are also unable to consume sea water for survival. That's because they retain their mammalian kidneys which cannot process salt water.

These are major threats to survival, yet the whales failed to evolve into truly marine creatures. They evolved a way to hold their breath longer than any other mammal , their 'nose' migrated to the top of the skull, but their internal organs, the lungs and kidneys, and their reproductive system have remained that of a mammal.

Interestingly like all mammals the embryo of the whale passes through a fishlike stage that has gills (the branchial arches) and 'primitive' fish like kidneys. Given that evolutionists use embryos as intermediates in explaining several evolutionary steps, a prematurely born whale embryo could have survived and given rise to a class of whales that could breathe through gills. That didn't happen in 50 million years.

If convergent evolution were to occur across physiologic boundaries each habitat would be populated with animals with identical morphology, physiology and reproductive system best suited for the environment.

These would be whale-shark-turtle hybrids.

Mammalian brain: quick to learn, streamlined shape of a shark and a turtle like armour.

Misfits in water *: The Turtle and the whale compete with the shark in its natural habitat yet they retain their native physiology. Both the turtle and the whale are unable to breath underwater. The turtle can't lay eggs in the water whereas the whale cannot consume sea water. These are significant handicaps that should drive evolution*

Convergent evolution is limited to morphologic changes. Evolutionary process does not allow transformation of organs nor does it allow transformation of the reproductive system.

True convergence*: If evolution were to occur across physiologic boundaries a shark like creature with mammalian brain and a turtle like armour would have emerged through convergent evolution*

There are other examples. The transformation of amphibians to reptiles.

Frogs have inhabited earth for at least 200 million years. There are few amphibians that make the desert their home. There are desert toads. They survive by digging deep holes in the desert and hibernating for the most part of the year, and seek temporary water bodies to reproduce. Several frogs allow the tadpoles to grow in their skin or carry them on their moist backs. None of the adaptations have resulted in transformation of the reproductive system to a system of hard eggs, independent of water which would allow them to thrive in arid climates.

Thus beta transformation also fails when examined for viability of intermediates

GAMMA TRANSFORMATION: EMERGENCE OF NEW FORMS WITHIN MAMMALS (ORIGIN OF ORDERS)

Mammals exist in a massive array of forms; rodents, bats, carnivores, herbivores, whales, apes and elephants. They all have descended from a primitive mammalian ancestor and gradually adapted to all forms of habitats.

This process is very much feasible through the process of Darwinian evolution.

Mammals gave rise to mammals that adapted to various available niches and developed widely different external forms. Monkeys, rodents, whales, carnivores and the ungulates.

These evolutionary changes however seem to lack universal uniformity across animals that share the same environment and share the same threats to survival.

Whales look like fishes due to evolution for aquatic environment; however seals, and sea lions, sea-cows, somehow never acquired the same fishlike form.

Similarly, for the adaptations seen in the giraffe. It seems that the long neck evolved from short necked ancestors in order to exploit the inaccessible leaves on tall trees. Several species of antelopes have shared the same ecosystem and faced the same environmental threats. They could have benefited from a long neck, yet they did not converge into long necked antelopes in the very same environment.

Antelopes and deer have short ineffective horns and small bodies that make then vulnerable to predators. Yet none have developed rhino like bodies and an impenetrable hide; something that would enhance their survival against predators.

Predators such as the lions could have benefited from large size and could have easily hunted large prey such as elephants for meals, yet they have remained comparatively small even though they were descendants of, the sabre tooth tigers that were much larger in size.

Gamma 2: **Origin of families**: separation of dog and cat family within the carnivores. There is moderate degree of morphologic transformation with no changes in physiology or reproductive systems. This form of evolution is feasible from the point of view of survival as well as reproduction.

Gamma 3: **Origin of species**: occurs within the family. Separation of dogs from wolves (which represent various genera in taxonomic terms). Again highly feasible. These groups have minimal morphologic differences and may sometimes interbreed. It is feasible on account of survival and reproduction, and is easily observable.

Incidentally all of Darwin's examples of pigeons, sheep, and finches arising from single ancestral species fall in this group.

In summary the biological feasibility model reveals;

Alpha transformation (evolution of multicellular animals from unicellular animals) was biologically unfeasible

Beta transformation (evolution involving transformation of reproductive system or basic physiology) was biologically unfeasible

Gamma transformation: Transformations within the same class: was Biologically Feasible

Gamma2 Transformation: was Biologically Feasible

Gamma3 transformation: was Biologically Feasible and observable.

The conclusions are summarised in the table on the following page

Level	Evolutionary step	Biologic basis of transformation	Biological feasibility
Gamma 3	Origin of species	Small phenotypic transformation (wolves/foxes/dogs)	**Feasible**
Gamma 2	Origin of families	Moderate phenotypic transformation (cat vs dog family)	**Feasible**
Gamma Transformation	Origin of orders	Major phenotypic transformations without transformation of reproductive system (primates/bats/rodents within mammals	**Feasible**
Beta Transformation	Origin of classes	Transformation of reproductive system and organs	**Unfeasible**
Alpha Transformation	Origin of Phyla Fist metazoan from unicellular organisms	Origin of new body plans from unicellular cells including organs and reproductive system	**Unfeasible**

Summary of biological transformations and their feasibility based on the criteria of survival and fertility.

WHAT OF THE SCIENTIFIC EVIDENCE: WHERE DOES IT FIT?

Scientific literature is full of evidence that supports the theory of evolution.

We cannot ignore this fact; however, with the new model of evolution we can place the evidence in the right context.

Darwin's Finches and the Peppered moth: Darwin's iconic example of finches is fundamentally speciation from ancestral Finch based on the shape of the beak. It falls in the level of $\gamma 3$ transformation, the smallest level of transformation along with the other icon of evolution: the Peppered Moth, where moths with black spots on white wings evolved into all black moths in response to blackening of tree trunks from soot during the industrial revolution. This is akin the emergence of subspecies of dogs form ancestral dogs.

Homologues: Homologous structures are considered an example of evolution. The forelimb of the horse, the forelimb of man, the flipper of whale and the wing of the bat have similar bones. The structures point to similarities in skeletons within the class of mammal (γ **transformation**). There are similarities to wings of birds and upper limbs of reptiles and amphibians. These similarities remain morphologic similarities and do not explain the transition of reproductive mechanisms (β transformation) or internal organs such as the kidneys.

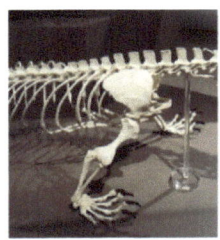   

REPTILE   MAMMAL (QUADRUPED)   DOLPHIN   WHALE

Similar (homologous) upper limb skeleton in various groups of animals

Vestiges: Vestigial structures are structures present in different animals as rudiments of well-developed structures in others. Common examples are the limbs in snakes, and tails in humans. These represent morphologic transformations of existing structures in the process of evolution within the same class of animals (γ transformation).

The shrunken limbs in snakes indicate four legged reptilian ancestors, whereas the shrunken tail in humans indicates ancestral primates with tails, all within the same class (reptile and mammal respectively)

Convergent evolution: The body shape of dolphins and whales has evolved to resemble fish suited for swimming and is considered an evidence of evolution. It is true, however the convergence is purely morphologic, without the transformation of mammalian physiology or the reproductive system .These changes again remain within the confines of the class **(γ transformation).**

It also underscores the fundamental limitation of evolution; that evolution cannot occur across physiologic boundaries and cannot change reproductive systems.

Embryology: During its development the mammalian embryo passes through a stage resembling fish anatomy followed by an amphibian anatomy, then a reptilian anatomy and finally shows mammalian anatomy. This was taken as a support for evolution (ontogeny repeats phylogeny).

As we have seen earlier, although transition from one reproductive system to the other can be observed in an embryo; it is not possible in evolution where the intermediates that have transitional forms of reproductive organs need to reproduce to evolve to the next stage of development.

Last but not the least one needs to remember that the embryo has a complete genetic blueprint to progress through various stages of development to a final form, whereas evolution is an unguided process.

Genetics: Cells of all living animals contain DNA. DNA contains the programme for building the entire organism.

In a simple way the genetic information is coded in specific genes that direct synthesis of individual proteins.

We can state that multicellular organisms would share several genes present in unicellular organisms. The ciliated cell of a sponge is likely to share genes which program cilia in unicellular organisms, hence the sharing of 230 genes. At the level of tissues animals are similar across the phyla. Muscle cells, nerve cells, secretory cells, bear uncanny similarities.

Presence of similar genes indicates the same fundamental building blocks of biology, it is a proof that all organisms follow the same biology. They are made of the same proteins and molecules, however biological relatedness does not prove that they descended from one another and that transformation from one form to the next was biologically feasible.

The following illustration pairs the levels of transformation with the available scientific evidence:

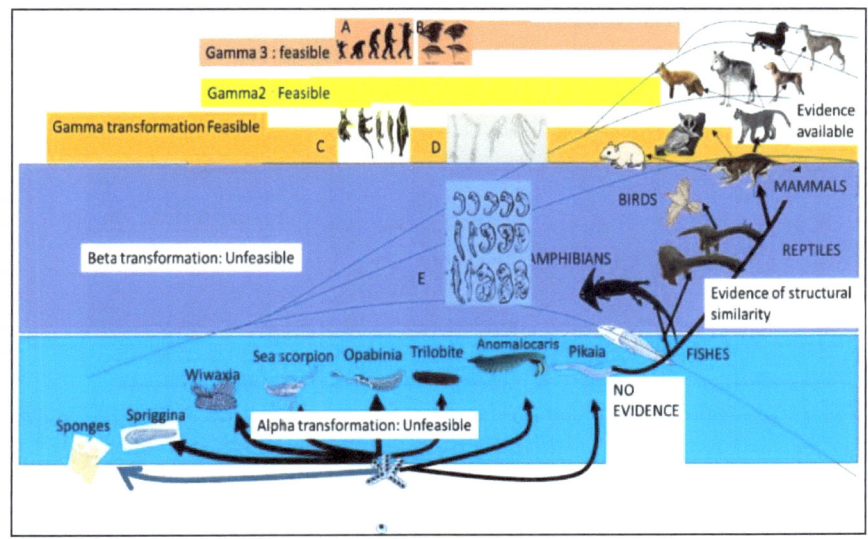

Available evidence and its status according to the levels of transformation

A: Evolution of primates: *γ3 transformation among primates Biologically feasible**: Finches**: γ3 transformation amongst finches: Biologically feasible.* ***C: Evolution of whales***: *γ transformation among mammals (biologically feasible)* ***D: Homologues*** *and vestigial organs: γ **transformation** among mammals/reptiles (Biologically feasible)* ***E: Embryology*** *is used to support **β transformation**. Evidence of similarities in **skeletal structure** is seen for transformation between reptiles and mammals however transformation of reproductive system between amphibians to reptiles and reptiles to mammals is inexplicable through Darwinian evolution **Evidence of beta transformation***: *examples of turtles and whales demonstrates that evolutionary transformation does not occur across reproductive systems. **The evidence of alpha transformation** is missing and is purely hypothetical.*

COMPUTER MODELLING: THE BLIND WATCHMAKER: AN ERRONEOUS PARADIGM?

Dr Richard Dawkins is one of the most public proponents of evolutionary theory. In order to explain evolution he proposed the Blind watchmaker model. He hypothesised that evolution is like a blind watchmaker. It does not have a purpose or direction.

It was a pun to refute the watchmaker analogy by William Paley in his 1802 book Natural Theology.

Paley stated that the complexity of living creatures was itself an evidence of the existence of a divine creator as the existence of a watch compels belief in an intelligent watchmaker.

But, it may appear that, unknowingly Mr Dawkins proved the very point which he was trying to refute.

Dawkins illustrated the difference between the "potential for the development of complexity" through randomness, "as opposed to that of randomness coupled with cumulative selection".

He created a computer model of artificial selection implemented in a program also called The Blind Watchmaker.

"The program displayed two dimensional shapes (a "biomorph") made up of straight black lines, the length, position, and angle of which were defined by a simple set of rules and instructions (analogous to a genome). Adding new lines (or removing them) based on these rules offered a discrete set of possible new shapes (mutations), which were displayed on screen so that the user could choose between them. The chosen mutation would then be the basis for another generation of biomorph mutants to be chosen from, and so on. Thus, the user, by selection, could steer the evolution of biomorphs.

This process often produced images which were reminiscent of real organisms for instance beetles, bats, or trees. Dawkins speculated that the unnatural selection role played by the user in this program could be replaced by a more natural agent if, for example, colourful biomorphs could be selected by butterflies or other insects, via a touch sensitive display set up in a garden".

The important issue that was overlooked, was the computer program itself.

The computer program was created by an intelligent agent (Mr Dawkins himself or the coder) which took instructions from external sources and drove evolution.

The computer program could not have been the product of a random process by itself.

The computer program represents the genetic blueprint, which gave rise to the very first living animal. It is the prerequisite of evolution and not the product of evolution.

This model inadvertently proves the existence of intelligent design rather than refute the point.

Besides the computer program*, again the biological feasibility of early and incomplete forms was ignored*. Early and incomplete forms of biomorphs would neither survive nor reproduce

HOW DOES BIOLOGICAL FEASIBILITY MODEL COMPARE WITH SCIENTIFIC OBSERVATIONS

To be considered true, the predicted pattern of evolution using the biological feasibility model should correspond to the actual observed pattern of evolution and the available evidence.

Alpha Transformation: we predicted Alpha transformation, (the transformation of unicellular organisms into multicellular animals) was not biologically feasible.

It is well known that there is no tangible evidence of this transformation. The gap is filled with several hypotheses with inclusion of hypothetical or imaginary intermediates that have been proposed to follow the evolutionary framework. We also observed that the hypothesis was faulty.

This transformation corresponds to the inexplicability of the Cambrian explosion where representatives of all major phyla of multicellular animals appeared simultaneously.

Beta transformation: We had predicted that evolutionary steps where evolution involves transformation of reproductive system, or internal organs, (unique to the physiology and body plan) would not be biologically feasible.

This conclusion corresponds to the observed evolution.

It is a scientific fact that once the initial phyla had come into existence during the Cambrian period they have remained constant. The original phyla have not given rise to any new phyla and new members have stayed within the boundaries of the original body-plan.

Evolutionists call it the mystery of ***the fixity of phyla.***

A similar scenario is seen in the evolution of vertebrates, where the evolutionary theory suggests that amphibians evolved into reptiles which then gave rise to mammals. However this conclusion is based on skeletal similarities between the groups. In order to prove this transition it would be imperative to demonstrate the existence of an intermediate form of reproductive system, i.e. a reproductive system that would have features of a reptile and the mammal. Till date No fossil has been shown to have an intermediate form of reproductive system that would represent the transition, nor such a reproductive system theoretically conceivable.

We also observed that amphibians, reptiles and mammals have adapted to extreme environments over several million years yet have not been able to transform their reproductive systems or their native organs. They have all maintained their native physiology and adapted within the limits of the physiology.

This can be termed as the *fixity of classes*.

Gamma transformation: evolution of orders within the same class of animals was deemed feasible through the process of natural selection and some, (but not all) transformations can be observed in fossil records. It includes evolution of modern reptiles from ancestral reptiles. Modern mammals from prehistoric mammals and birds from birds. These evolutionary transformations are feasible and several of them can be observed in fossils these include evolution of whales, evolution of horses, elephants and primates

Gamma2 transformations: This is the evolution of families within each order such as the separation of the cat family from the dog family. These evolutionary transformations are feasible and also seen in fossil records.

Gamma3 transformations: are the evolution of new species form existing species. These highly feasible are observed in fossil evidence as well as observable. They are observable in nature as well as in artificial breeding

The summary is seen in a tabulated form on the next page:

Transformation	Biologic basis of transformation	Biological feasibility	Observable Evidence
Gamma 3	Small phenotypic transformation (wolves/foxes/Dogs)	**Feasible**	**present**
Gamma 2	Moderate phenotypic transformation(cat vs dog family)	**Feasible**	**present**
Gamma	Major phenotypic transformations **without transformation of reproductive system** (primates/bats/rodents within mammals	**Feasible**	**present**
Beta	Transformation of reproductive system and organs	Unfeasible	Absent (Evidence is based on structural similarities)
Alpha Transformation	Origin of new body plans from unicellular cells including organs and reproductive system	Unfeasible	Absent. Purely hypothetical

A summary shows that absence of physical evidence corresponds to the biological feasibility of various transformations.

Pitting evolutionary steps against the conditions of biological feasibility demonstrated that the very first step in the evolution, transformation of unicellular animals into multicellular organisms, was not biologically possible through the step wise process of natural selection.

Similarly transformations amongst multicellular animals that involved replacement of internal organs or the reproductive organs were not possible.

The following illustrations show the comparison between Darwinian model, the actual fossil data and Bio-feasibility model

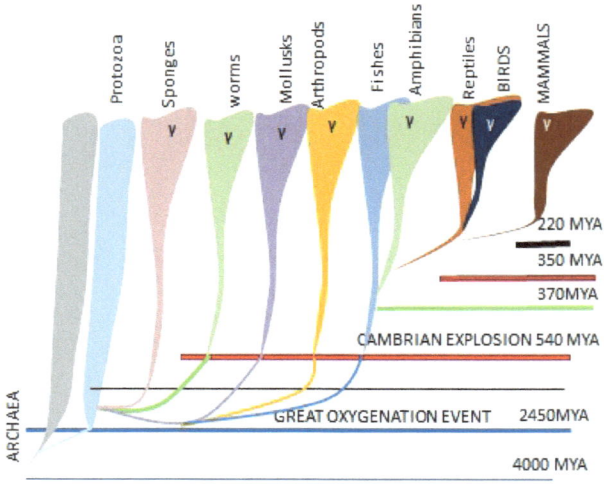

Current model of evolution based on the paradigm of common ancestry. Each phylum originating from a pre-existing phylum

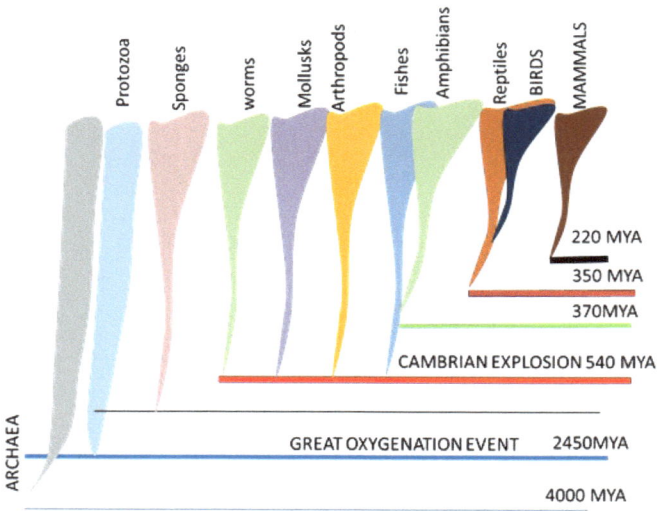

Actual fossil Data showing discrete origin of each lineage of animals, with no intermediates between groups

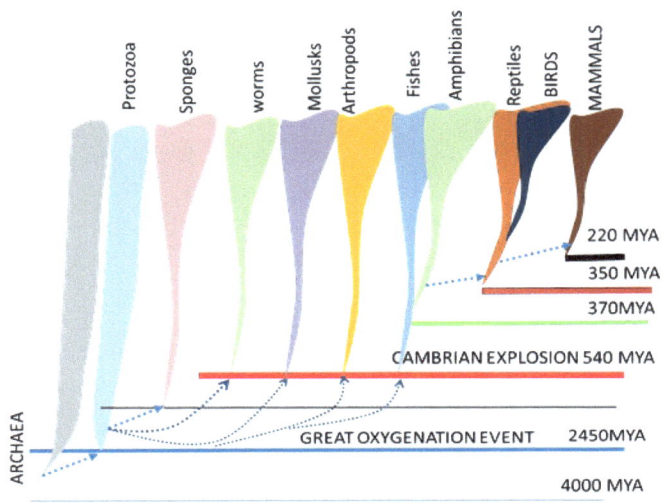

Dotted lines indicate the hypothetical intermediates based on the Darwinian paradigm, without actual fossil evidence.

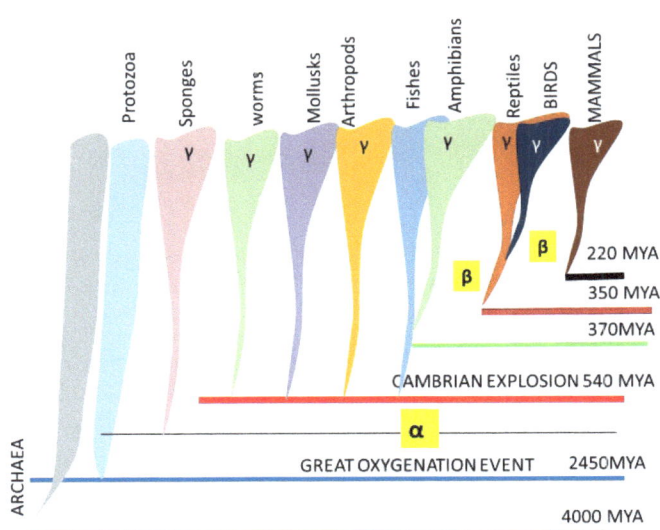

Biological feasibility model removes the feasibility of alpha (α) and beta (β) transformations, and corresponds to fossil data.

DARWIN'S ASSUMPTIONS

When Darwin wrote: 'the origin of species through natural selection' he was well aware of the ***term 'species' and his intention was to explain the origin of new species being from existing species***, This becomes quite clear from the examples he had cited, such as new species of pigeons from rock pigeons, or finches with different beak shapes from ancestral finches, or varieties of sheep from ancestral sheep.

All his observations fell within the realm of species or at the most the same genera. None of his observations crossed the class of animals (for example reptiles from amphibians or mammals from reptiles).

The mechanism for the origin of new species (within the scope of species or genera) was natural selection, which was derived from the idea of ***selective breeding*** seen in animal husbandry.

In other words ***natural selection was a form of selective breeding in nature*** brought on by environmental pressures, such that individuals with advantageous variations survived and progressively lead to emergence of new species from an existing population of species.

He wrote *"Owing to this struggle for life, any variation, however slight and from whatever cause proceeding, if it be in any degree profitable to an individual of any species, in its infinitely complex relations to other organic beings and to external nature, will tend to the preservation of that individual, and will generally be inherited by its offspring ... I have called this principle, by which each slight variation, if useful, is preserved, by the term of Natural Selection, in order to mark its relation to man's power of selection".*

Over time he began to extrapolate this concept to fossil records, until he came to believe what became a universal theory to explain the ***origin of all organisms*** well beyond the realm of 'origin of species'.

Darwin's idea of struggle for existence:

Darwin proposed the idea of struggle for existence as a driver of evolution. The population of an animal exceeded the natural resources leading to competition amongst the members of the species and the survival of the fittest.

The idea had stemmed from the theory proposed by Malthus.

In his 1798 writing '*An Essay on the Principle of Population*', Malthus observed that *an increase in a nation's food production improved the well-being of the populace, but the improvement was temporary because it led to population growth, which in turn restored the original per capita production level.*

Looking closely the statement actually does not represent a model for continuous improvement of a population through struggle. If fact it is just the opposite. It states that abundance of food allows populations to expand which neutralises the benefits of socioeconomic improvement and a new norm is reached with a lowering of socioeconomic status. In other words the resources get redistributed within the population and the population reaches a new equilibrium with the resources.

It is true that struggle for survival happens in nature. There is an intra-species and interspecies competition for resources, but we are also aware of the concept of *ecological balance*.

In any given ecosystem all animal populations including predators and prey stabilize at one point. There is never a continuous ever- escalating struggle or an arms race. In periods of scarcity big chunks of population perish and survivors start a new population.

We can see these cycles in several ecosystems that are affected by cyclical draught. An example would be cyclical droughts in African ecosystems. These droughts affect mammals such as antelopes, wildebeest, reptiles (such as crocodiles) and fishes in the rivers. Although survivors may be hardier that their counterparts, even after several thousand generations of similar cycles (over the last century), we do not observe adaptations that would make the offspring completely immune to draught at any point.

Secondly, when we look at primitive ecosystems, between cycles of ice age, the climatic conditions had remained stable for millions of years.

If we look at primitive oceans that harboured first life, they were stable for millions of years. Moreover one may imagine that at the dawn of animal life, the population of animals would have been small relative to the size of the habitats.

There were entire ecosystems of predators and prey which would balance the population, making it hard to imagine an overcrowding of animals in the primitive oceans driving the fish out of water from sheer struggle for survival.

What does survival of the fittest actually mean?

Evolution, by definition is modification with descent. It can only be applied to an existing population that is capable of survival and breeding over several generations in its given habitat. In essence it is a population of *fit individuals giving rise to fitter individuals, and not, individuals unfit to survive giving rise to individuals fit to survive. This is again a contradiction of terms*

This would be the situation when we speak of evolution of organs essential for life. Intermediates born with underdeveloped organs will simply die, and logically the cycle will continue endlessly. Survival would only be possible if an entire organism is formed in a single step with all organs fully developed and functioning.

This also applies to individuals born with rudimentary reproductive organs. In absence of reproduction, it would be logical that they would become extinct within the same generation. The process will continue endlessly, until the entire system comes into existence in a single step.

Essentially the emergence of a viable organism from stages of non-viable organisms or the emergence of a fertile organism from infertile stages is self-contradictory.

Thus transition from one celled organisms to multicellular organisms which was based on the development of new organs and new reproductive system, and the evolution of animals with new reproductive forms both could not have occurred through the process of natural selection.

This explains why all of the scientific evidence supports evolution within the same reproductive group, and the evidence is entirely hypothetical for the emergence of multicellular animals from single celled organisms.

It also points to the misconception that evidence gleaned from observing evolution within the species can be extrapolated to the entire process of evolution i.e. emergence of new phyla or new classes of animals that have different organs and reproductive systems.

The transformation of organs is possible through the process of natural selection:

Darwin had also noticed that animals of different groups had different organs and different physiology. He explained that the development of internal organs also followed the same principle and they evolved in stages.

He wrote:

*"If it could be demonstrated that any **complex organ existed, which could not possibly have been formed by numerous, successive, slight modifications**, **my theory would absolutely break down**. But I can find out no such case"*

The idea possibly came from observing embryos of vertebrates, where the organs transform from fish like stage to amphibian ,to reptilian and then mammalian stages.

From the discussion on biological feasibility, one can see the fallacy of transposing evidence from embryological development on evolution.

Internal organs can be seen to evolve in stages in an embryo, however the embryo can survive transitions through stages of immature or underdeveloped organs because it is supported by an egg or a placenta, *in the case of evolution however, intermediates with immature organs would need to survive and reproduce as independent creatures*, which would not be possible. Such intermediates would perish at birth and never be able to evolve further.

It is a common error to assume that the misfit species would become extinct and the fit intermediates would have survived and that also becomes the logical explanation for absence of fossils.

The truth is, that all misfit intermediates that do not have completely developed organs and reproductive systems would become extinct without progressing any further. Natural selection could have proceeded only if perfect and complete animals were formed at each step

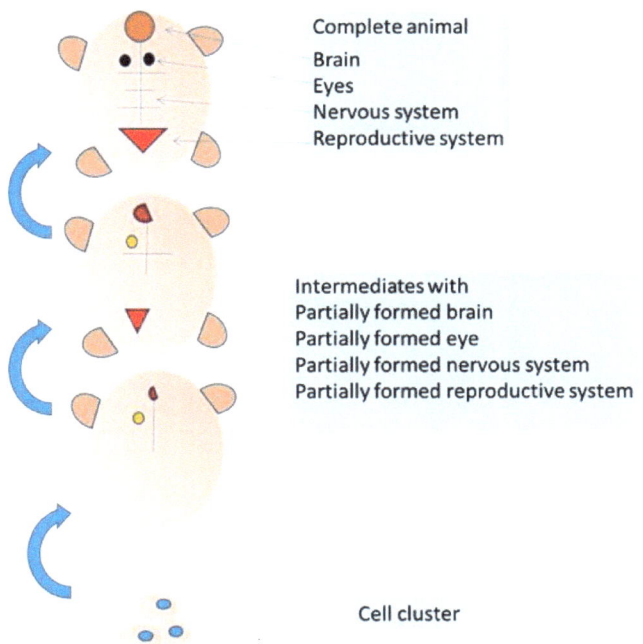

Darwin's assumption *that animals could develop complex through numerous successive slight modifications*

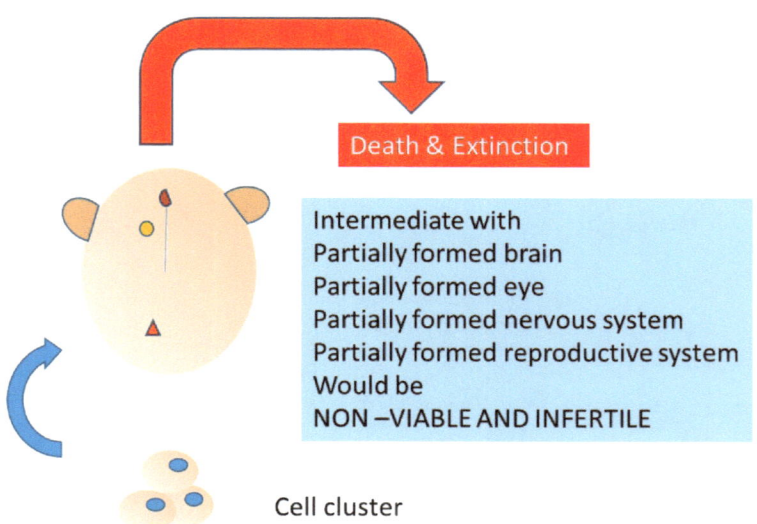

Actual fate of such intermediates: *animals with partially formed organs neither survive nor give rise to offspring*

LAYING DOWN RULES AROUND DARWINISM

Darwinian concept of natural selection has been allowed unlimited power and has been used to explain the entire process of evolution from a single cell to the whale. It is akin to infinite regression, where each population emerges from the previous one until a single cell bacterium remains. However based on our discussion some principles can be laid down.

Evolution by natural selection:

1. Cannot create or transform organs
2. Cannot create or transform the reproductive system
3. Can be applied only to evolution within the same class or phylum of animals where variations are limited to sizes and shapes or existing organs and body structures.

These principles explain several real life observations:

- The Cambrian explosion and the lack of intermediate fossils between one cell organisms to multicellular animals (such intermediates could not have existed)
- Fixity of phyla: Observed evolution has been restricted within the boundaries of each phylum. No new phyla have appeared since the Cambrian period, nor have any phyla merged.
- Why classes of vertebrates have remained distinct over millions of years even when they have adapted to similar environments. (turtles & whales)
- All existing fossil evidence demonstrates evolution within the same class of animals.

Natural selection applies only to morphologic variations observed within an interbreeding group of animals such as finches from existing finches, dogs from ancestral dogs or wolves, black moths from white moths. **It cannot cross boundaries of physiology and reproduction.**

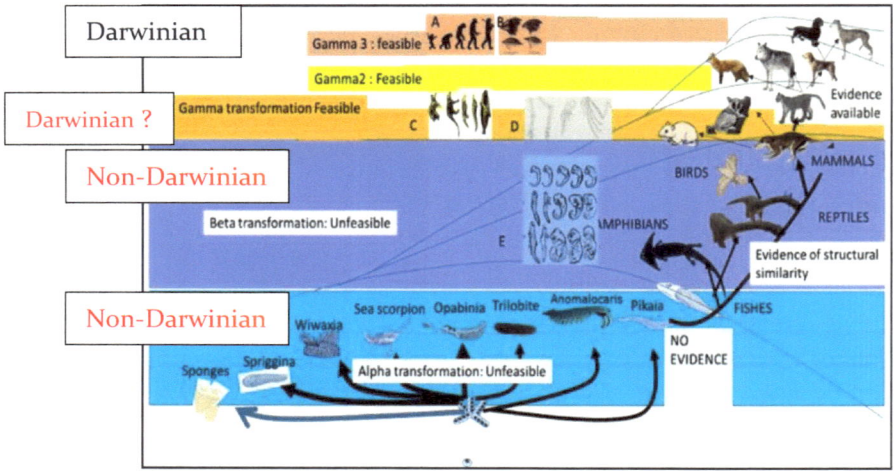

The transformations that are possible and impossible through Darwinian mechanism.

The transformation of single cell organisms to multicellular organisms includes generation of new organs and new reproductive systems, which is biologically unfeasible through natural selection,

For the same reason the phyla remain fixed and evolution through natural selection remains within the confines of the phylum.

The differentiation of classes (amphibians to reptiles and reptiles to mammals) is also biologically unfeasible through the process of natural selection due to major transformation in organs (such as the urinary system) and the reproductive system, however transformation from reptiles to birds is feasible due to the identical nature of the reproductive systems.

Evolution of new forms within mammals or within reptiles is feasible through natural selection and natural selection remains within the confines of the physiology of each class,

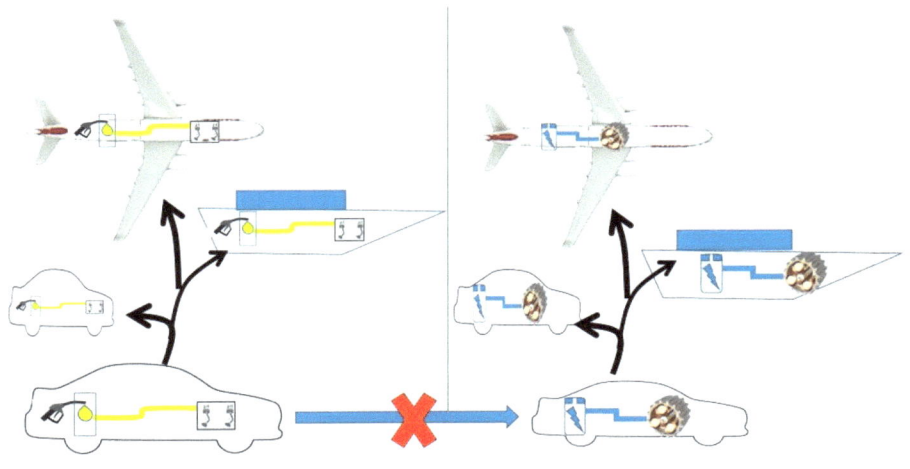

Gasoline class   *electric class*

Summary of fundamental rules governing evolution by natural selection:

A gasoline car (left) may evolve to a different shape and size, a gasoline powered boat or a plane depending on the habitat, as the changes are morphological without the change in the inherent physiology.

Similarly an electric car (right) may evolve into a smaller electric car or an electric boat or plane depending on the habitat,

However a gasoline car cannot evolve into an electric car (or vice versa)through the process of evolution as it involves transformation of its inherent physiology

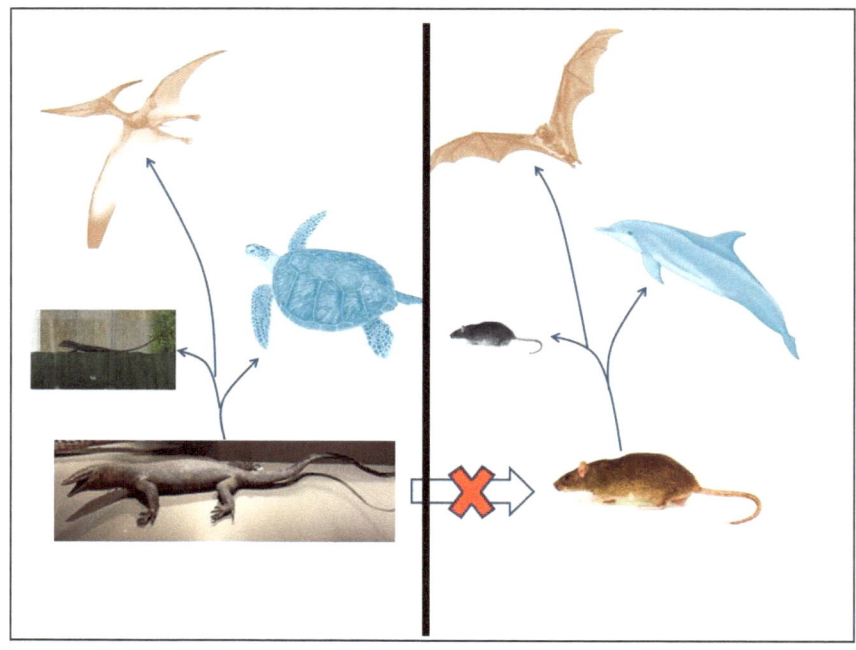

Evolution by natural selection is biologically feasible within the same class of animals, with the same reproductive system and the same physiology. It cannot cross these boundaries.

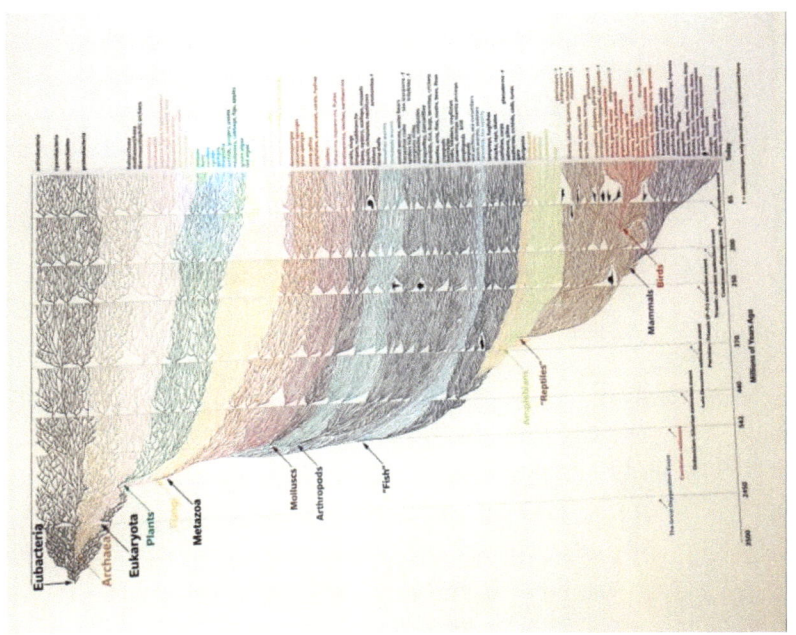

The modern evolutionary history follows the same model. *In its entirety Evolution strictly stays within the boundaries of the individual physiology and reproductive anatomy of each lineage without crossing over.*

In spite of this glaring evidence, it is believed that each individual phylum arose from a single cell ancestor

SUMMARY

The theory of evolution was born out of one man's philosophic conclusions which were derived from the simple concept of selective breeding. The theory was intended to specifically describe the origin of 'species' as is evident from the title.

Darwin had observed the effects of artificial selection in breeding domesticated animals and postulated that in nature, new species emerged from a population of existing species through the process of natural selection as they struggled to survive, while accumulating advantageous variations.

This could easily explain emergence of several varieties of pigeons from rock pigeons, several breeds of dogs from wild dogs and several varieties of finches from one population of finches

At some point the concept was extrapolated to explain the emergence of all life on earth from a common ancestor leading to the establishment of the theory of evolution.

The entire theory rests upon one single assumption, that any degree of biological transformation was possible in living animals through accumulation of genetic changes over generations. One did not need to examine the feasibility of these transformations.

It was assumed that internal organs could be formed in stages in the same way as external morphologic features could evolve.

Darwin himself accepted "If it could be demonstrated that any complex organ existed, which could not possibly have been formed by numerous, successive, slight modifications, my theory would absolutely break down.

If we put this statement to test for its biological feasibility, the theory does breakdown.

Staged development of organs can be seen in embryology, however is not biologically feasible in evolution.

The fundamental difference is that during the stages of immaturity, while the organs are non-functional, the embryo is nurtured within the biological environment of the egg (or the uterus).

The embryo does not need to survive independently during these stages, nor does it need to reproduce to transition from one stage to another, which is the case in evolution.

In the process of evolution the intermediates with underdeveloped organs would fail to survive. Theoretically they would resemble aborted foetuses that are destined to perish.

A similar case can be made for the reproductive system. An organism with partially developed organs of reproduction would fail to reproduce and become extinct. Evolution of reproductive organs by itself is a contradiction of terms.

The limitations of evolution can be demonstrated using the biological feasibility model and the hypothetical intermediates evaluated for their ability to survive and to reproduce.

The biological feasibility model contradicts the assumption that all degrees of change could be explained through the mechanism of staged development through natural selection over generations.

The model predicts that natural selection cannot cross boundaries where replacement of internal organs and more importantly the reproductive system is involved.

This is the reason why phyla which represent unique body plans, have not changed or given rise to new phyla since the dawn of animal life 500 million years ago. Flat worms have remained flatworms, arthropods have remained arthropods and molluscs have remained molluscs.

It is the same reason reptiles have remained reptiles and mammals have remained mammals over millions of years irrespective of their habitats.

Turtles living in seas for 250 million years have not acquired gills, or evolved to reproduce underwater.

Whales and dolphins have inhabited seas for 50 million years have retained their mammalian lungs and cannot breathe underwater. They cannot even consume the sea water they swim in. They are unable to evolve these functions.

In both the cases we can see the failure of evolution even when these handicaps are a serious threat in their struggle for survival.

It is no surprise that evolutionary changes described by Darwin and fossil evidence seen today are limited to variations seen within the same group of animals. (Morphologic changes within the same species such as finches, or homologues organs seen within mammals)

We also know from past the 30,000 years of animal domestication and the past 500 years of scientific experimentation, that **no amount of selective breeding can bring about a change the fundamental physiology of an animal.**

Reptiles can never be bred to demonstrate mammalian characteristics or vice versa. Animals can only be bred to alter morphologic features within their own class.

The **simple concept of selective breeding or natural selection** as it was termed, was simply a tool that became available to the scientific community to provide a convenient explanation for the biodiversity with complete disregard for true scientific inquiry and **became the theory of Evolution**.

The limitations of the theory must be accepted and serious thought needs to be given for alternative explanations for the origin of biodiversity. Instead of hanging on to a faulty paradigm, the focus of science should shift to finding real answers.

POINTS TO PONDER

For just a few minutes, let's remove the glasses of evolution and take an honest look at biodiversity over the ages.

If we can conclude that multicellular life forms could not have evolved through the process of evolution, nor could have novel reproductive forms evolved through the process of evolution, alternative explanations need to be sought.

EACH ERA HAD AN ENTIRE ECOSYSTEM OF ANIMALS

It is also interesting to observe that in **each era, an entire ecosystem of plants and animals existed, in which several forms of each class of animals arose and existed simultaneously. All of them were fully formed and capable of survival.**

The Cambrian period had an entire ecosystem of invertebrates, in the Silurian period several types of fishes existed, during the age of reptiles an entire ecosystem of dinosaurs appeared simultaneously. During the age of mammals an entire ecosystem of hundreds of species of mammals existed simultaneously.

https://www.faraday.st-edmunds.cam.ac.uk/CIS/rolston/Rolston%20-%20lecture.htm,

Niklas, Karl J. 1986. "Large-Scale Changes in Animal and Plant Terrestrial Communities." Pages 383-405 in D. M. Raup and D. Jablonski, eds., Patterns and Processes in the History of Life. New York : Springer-Verlag.

Niklas, Karl J. 1997. The Evolutionary Biology of Plants. Chicago : University of Chicago Press.

If it were a question of evolution of all forms from a single species of each class, then we would find abundance of a species that would represent the entire class, which would give rise to the next group of species a million year later and so on and so forth. It would take millions of years to evolve just a few species, and literally a billion years to build a single ecosystem.

Absence of imperfect forms

Every fossil ever seen shows a completely formed organism. There are no partially formed or malformed animals struggling to become perfect. Even if one argues that imperfect forms perished, we would have at least found some fossils of malformed animals.

A primitive fish fossil shows a body perfect for swimming with skeletons and joints in perfect shape and order, insects such as dragonflies and ants of 35 million years ago hardly show any difference from their counterparts today.

The Tyrannosaurus *rex* can be considered a far more advanced, compared to any modern day reptile. Each species lasted several million years. Ancient invertebrates such as sponges, insects, and nautiluses continue to exist unchanged.

One may place these animals on a chronological map and state that each new form evolved from the previous form, however it is simply the case of one ecosystem of animals replaced by another.

What is primitive? Is there any organism that is imperfect? A fossil fish showing a complete skeleton similar to modern fish. Cambrian fossils sow perfectly formed trilobites and mollusc shells, similar to modern arthropods and molluscs respectively.

A SERIES OF FORTUNATE EVENTS, OR PLANNED 'TERRAFORMING' OF EARTH?

The extraordinary orderliness and happy coincidences points to the possibility of **planetary engineering** or terraforming with sequential preparation of the planet and introduction of life-forms with an evolving ecosystem.

The position of the earth in the solar system "the Goldilocks zone" which could sustain liquid water and temperatures just right for life (a happy coincidence).

Bio-chemo transformation of the planet: If we go back to the first events of this planet, the environment was a pool of boiling chemicals, devoid of oxygen and water. We believe it was transformed to a cradle of life by that accidental appearance of bacteria that could thrive in these boiling temperatures (where most life would perish), loved to feed on these chemicals and knew how release oxygen and water into the environment. This is known as the Great oxygenation event.

Was it **a happy coincidence** or a purposeful step at terraforming with a deliberate introduction of chemo-thermophilic bacteria to transform chemicals into water and oxygen to prepare the planet for the introduction of oxygen based life?

Once water and oxygen levels had stabilised unicellular animals and plants 'appeared' (or were they introduced) to create a microbial ecosystem.

At the next stage we see the appearance of filter feeders like the Charnia, that were equipped with a feeding mechanism just right to filter microbes from water, to create a simple ecosystem along with small animals that could feed on the sponges.

Once established this was followed by the Cambrian explosion and the appearance of numerous aquatic invertebrates perfectly suited for the aquatic environment. Some of them were predators and others were the prey, creating a full aquatic ecosystem.

Jawless fishes then appeared that were perfectly designed to feed on soft prey. Fierce jawed fishes appeared soon, that could feed on the hard shelled arthropods and jawless fishes with skeletons.

Once the sea levels began to recede, amphibians appeared to inhabit the transitional zone between the water and land. Once land plants started proliferating insects seem to have appeared simultaneously for pollination, creating a terrestrial ecosystem. Were they introduced too? (As opposed to the hypothesis of co-evolution).

This was followed by a massive growth of vegetation on land. It required massive herbivores and equally massive carnivores to create a balanced ecosystem. However non- selective eaters were suitable, which also appeared simultaneously and a terrestrial ecosystem was created.

Increasing variety of plants appeared and animals with higher intelligence and learning abilities (birds and mammals) were introduced for selective feeding and pollination. The ecosystem now depended on behavioural and learning patterns of birds and animals.

Mammalian megafauna appeared such that the herbivores and carnivores were in proportion to the size and variety of vegetation.

With the appearance of humans an entire ecosystem of megafauna (Mammoths, Sabre tooth tigers, giant carnivorous birds) miraculously disappeared giving rise to an ecosystem of smaller modern animals. Extinction of megafauna has been attributed to human hunting; however primitive humans living in isolated tribes who could barely kill one mammoth in a week could not have been capable of eliminating an entire race of animals from the face of the planet

It has recently come to light that all modern fauna and humans are of the same age. Modern animals appeared simultaneously along with humans one begins to wonder, was it on a Noah's ark or the ark was a floating biosphere designed to repopulate the planet after the flood.

https://evolutionnews.org/2018/06/humans-and-animals-are-mostly-the-same-age/

Could all these events be planned by an intelligent race and the life synthesised in a lab, and introduced over time to set stage for a civilisation?

https://www.independent.co.uk/news/science/planet-born-formation-study-latest-image-picture-telescope-a8426451.html

ABOUT THE AUTHOR

Dr Amitabha Lahiri an avid researcher. His appreciation of the complex anatomy of human body and foray into the field of tissue engineering led him to question the oversimplified idea of evolution. This book represents the compilation of extensive research and insights gained over a period of almost decade of research on evolution and Darwinism.

Email: novumlibrum@gmail.com
Blog: https://evolutionconceptlimits.blogspot.com/

ACKNOWLEDGEMENTS

Heartfelt thanks to Dr. Manjari Lahiri for correcting the manuscript, and above all, believing in the idea!

Thanks to my children, Anushka and Sanjana, for holding me to the highest standards, teaching me how to be happy!

Special thanks to Sanjana for lending her beautiful artwork for the cover.

www.ingramcontent.com/pod-product-compliance
Lightning Source LLC
Chambersburg PA
CBHW040219220526
45473CB00001B/46